Biological Sciences

A Kaleidoscopic View of New Trends

About the Authors

Prof H R Tyagi

Prof. H. R. Tyagi born in 1944 and he have expertization in the area of Herpetology, cell biology, physiology of vision and environmental science. He has act as Head department of Zoology, college of Science, MLS University two times from 1986 to 1989 and 1996 to 1998.He act as officiating Vice chancellor and novel office for NAAC and several other bodies.

Dr. Ram Prakash Saran

Dr. Ram Prakash saran is presently working as Assistant Professor in Zoology at Jai Narain Vyas University Jodhpur (Rajasthan). He did his M.Sc. in Zoology from Jai Narain Vyas University Jodhpur in 2005 and Ph.D. in 2010. He was throughout first class in his studies up to M.Sc. He has cleared many national level examinations like UGC- CSIR NET, GATE Lifesciences, SET Lifesciences. He engaged in research on threatened bird species of India with special emphasis on vultures, cytotoxity of veterinary drugs on avifauna and trophic level transfer of nanoparticle and pollutants. At present he is involved in teaching diversity among animal forms, Animal behaviour and biological evolution. He has published more than 23 research articles and reviews on animal behaviour, and more than 10 books in national and international publication and peer reviewed journals.

Dr. Heera Ram

Dr. Heera Ram is currently working as Assistant Professor in department of Zoology, Jai Narain Vyas University, Jodhpur (Rajasthan), India. He did his postgraduation from Mohan Lal Sukhadiya University, Udaipur and doctoral degree from Jai Narain Vyas University in the field of atherosclerosis and therapeutic applications of desertic medicinal plants. He has published around 30 research articles in national and international journals and supervised three doctoral degrees. His research area focused on applicability of the traditional knowledge and herbal formulations in ameliorations in metabolic syndromes. Nowadays, some affords are also concern on the applicability of phytocompounds in diabetes associated complications and hypercholesterolemia complications.

Biological Sciences
A Kaleidoscopic View of New Trends

Prof H R Tyagi
Former Dean and Head,
Retd. Professor Dept. of Zoology
Mohan Lal Sukhodia University
Udaipur (Rajasthan)

Dr. Ram Prakash Saran
Assistant Professor in Zoology
Jai Narain Vyas University,
Jodhpur (Rajasthan)

Dr. Heera Ram
Assistant Professor in Zoology
Jai Narain Vyas University,
Jodhpur (Rajasthan)

SCIENTIFIC PUBLISHERS

Published by
SCIENTIFIC PUBLISHERS (INDIA)
5 A, New Pali Road, P.O. Box 91
Jodhpur 342 001 (INDIA)
E-mail: info@scientificpub.com
Website: http://www.scientificpubonline.com

ISBN: 978-93-92590-69-6 (HB)
ISBN: 978-93-92590-75-7 (EBook)

Printed in India

Festschrift in Honour of Our Reverend PROF. ASHOK K. PUROHIT

A Teacher affects eternity,
he can never tell where
his influence stops.

ALL HIS LOVING STUDENTS

Preface

Biological Sciences dealing in animals and their interactions with each other as well as with their environment is possibly the most significant of all the sciences as it forms basis of resource education. Tremendous transformations in the knowledge, tools, techniques and applications of biological sciences have led to remarkable advancement in terms of research and education dissemination. The adaptability of animal or living beings depend on successive interaction with surrounding environment where physiology and environment stand with its uniqueness. The physiology of individual animal or group of animal represents its capability of survival whereas the surrounding environment indirectly interfere in functioning of systems. This pursuit of comprehending the diversity and complexity in the functioning of ecosystems and organisms residing therein is never ending. The present book "*Biological Sciences: A Kaleidoscopic View of New Trends*" is a paradigm in the trajectory of exploration and erudition of nature. The book covers a vast area of research in the arena of biological sciences such as Wildlife, Ethnobotany, Ethnopharmacology, Reproductive Biology, Endocrinology, Animal Physiology, Environmental Impact Assessment, and some recent issue of environment respectively.

We hope that research articles anthologized in the book will be valuable for the students of science, research scholars and academicians having an interest in biological sciences.

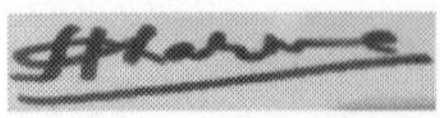

Prof. Kuldeep Krishan Sharma

(Former Dean Life Sciences and Head,
Department of Zoology,
University of Jammu,
Jammu -180006)

Contents

Contributers

C. Srinivasulu

Wildlife Biology & Taxonomy Lab, Department of Zoology, University College of Science, Osmania University, Hyderabad - 500 007, Telangana State, India & Centre for Biodiversity & Conservation Studies, Osmania University, Hyderabad - 500007, Telangana State, India

S. D. Kholkute

Former Director-in-Charge ICMR- National Institute of Traditional Medicine, Nehru Nagar, Belgavi-590010 and Former Director ICMR-National Institute for Research in Reproductive Health, Parel, Mumbai-400012

Sharma K. K

University of Jammu

Kantilal Hiralal Nagare

Department of Zoology, Birla college of Arts, Science and Commerce Birla College road, Kalyan- 421304

S.C. Joshi

University of Rajasthan, Jaipur (Rajasthan) India.

Sanjeev Kumar

DRC, ZSI, Jodhpur

H Ram

Assistant Professor, Dept Of Zoology, J.N.V. University Jodhpur

R.P. Saran

Assistant Professor, Dept Of Zoology, J.N.V. University Jodhpur

CHAPTER

1

Herbal Medicine: Challenges and Research Needs

S. D. Kholkute
Former Director-in-Charge
ICMR- National Institute of Traditional Medicine,
Nehru Nagar, Belgavi-590010
and
Former Director
ICMR-National Institute for Research in Reproductive Health,
Parel, Mumbai-400012

ABSTRACT

Herb and plant products are used as medicine for time immemorable. They are still in use for medicinal values and for treatment of various diseases. However, there is urgent need to carry out systematic and detailed studies to give traditional/herbal medicine due credit in global scenario. The article discusses various challenges and research needs in the area of herbal medicine.

Key words: Herbal medicine, pharmacology, quality control, standardization, toxicity, research needs, herb-herb-drug interactions

INTRODUCTION

"Herbal Medicine" includes all those medicinal products, which are made from plants, plant parts or any plant product prepared with minimum or no chemical manipulation (Iqbal 2018). Thus herbal medicine includes herbs / herbal materials like leaves, flowers, fruits, seeds, stems, barks, roots or any other plant parts. This also includes finished products that contains plants, plant parts etc.

Apart from primary metabolites, which are directly related in process of plant development, growth and reproduction; a large number of secondary metabolites are also synthesized from primary metabolites *e.g.* phenols, terpenes and alkaloids which are also useful for humans due to their medicinal properties. Thus their presence in any plant makes that as a medicinal plant with therapeutic activity.

Herbal medicine is one of the oldest form of health care. About 80% of world population still rely on herbal/traditional medicine especially for primary health care. About 30% of worldwide sales of drugs are based on natural products. Furthermore, herbal medicine has provided mankind with powerful drugs like morphine, quinine, cocaine, reserpine etc. to name the few. They also provide new lead structure/templates/scaffolds for drug discovery. India has tremendous wealth of plants with 47000 different plant species, almost 1/3 rd are designated as "medicinal." About 1500 medicinal plants are mentioned in ancient Indian literature. Though not successful so far in terms of medicine, natural products/medicinal plants are very important in drug discovery and development. They provide leads/pathfinder compounds with a greater structural diversity which are capable of being absorbed and metabolized. However, in spite of this only <10% world biodiversity has been screened so far for biological activities.

In recent years, there is resurgence in herbal medicine due to various reasons like: side effects of modern drugs; problems with drug resistant microorganisms, emerging diseases like chikungunyas, dengue where there are no suitable allopathic medicines are available. Furthermore, there is a great need for effective medicines for lifestyle diseases eg. cardiovascular diseases, diabetes, arthritis etc. They may also serve to improve quality of life eg. Rasayana.

There are certain positive features of herbal medicine like availability/accessibility, socioeconomical and culture aspects, cost effective, and presumed lesser side effects. However, there are some constraints which are affecting their global acceptability. The purpose of this article is to highlight these challenges which hinder wider acceptability of herbal medicine.

Authentication and correct identification of medicinal plants

Use of wrong plants or a wrong plant parts may be ineffective or even harmful. Therefore, authentication of plant material right from harvesting, collection, storage to finished product is essential and crucial. A variety of taxonomic, macroscopic, microscopic and phytochemical methods are already available. In recent years, various genomic, proteomic and molecular methods are used for correct identification of herbal drug component (Hegde *et al.,* 2018, 2019). Furthermore, techniques like diffuse reflectance (DRIFT) and Photo-acoustic spectroscopy (PAS) has also been recommended.

Substitution/adulteration is another issue of concern. Plant drugs have been described by their commercial or common/vernacular names rather than botanical names. This causes confusion in correct identification and use of plant material. Five different species are available in the market under the name Shankhapushpi, 6 species under the name Ratanjot, and some 10 species as Pashanbhedi. All these species resemble morphologically to the respective drugs as described in ancient literature. On the other hand, a single plant species is also used under different names. Furthermore there is lack of certification (passport) data of raw material. Thus, the correct botanical identity of so many 'single herbal drugs' is still confusing and controversial. The correct identification of plant may be also by the level of their active ingredients and marker compounds (phytochemistry approach). However, difficulty here is, that amount of active ingredients may vary with habitat, depends on soil, and climatic conditions and season of collection of raw material. Characterization of single herbal drugs is usually done by microscopic, macroscopic and routine chemical methods. This will be highlighted in quality control and standardization.

Quality control of herbal products

Quality and stability of herbal formulations have been neglected aspects. However, during the last decade there is a considerate improvement in this area. Standardization is adjustment of given contents of phytoconstituents/chemical markers. There is an urgent need of standardization as herbal products are often complex mixtures and active ingredient may not be known. Furthermore, dried parts are commercially available and procured, hence cross contamination and mis-identification is possible. Active ingredients may vary as per the plant part used, season of collection, climate and soil as already stated. Hence, quality control and standardization is a must to ensure uniform potency, to avoid batch to batch variations, for predictable efficacy and for comparison with known drug. Various stages of standardization of herbal medicine are:

- Identification and authentication of species. This should be done by expert plant taxonomist. A voucher specimen should be preserved and also deposited in well recognized laboratory like NITM, Belgavi.
- Physicochemical evaluations like ash value, extractive value, moisture content etc.
- Chemical evaluations consisting of identification and determining contents of marker compounds as mentioned in Pharmacopea.
- Microbial contamination should be under specified limits. Similarly, pesticide residue, heavy metal content should also be determined if the material is to be exported.
- DNA fingerprinting and state of the art techniques like DRIFT and PAS have also been recommended in case of controversy or confusion.

Safety and toxicology

This is very important and neglected aspects of herbal medicine. Extensive research needs to be undertaken if we wish to make plant drugs globally accepted. Sometimes reverse pharmacology approach (from clinical to laboratory and back to clinic) is also suggested as recommended by Patwardhan *et al.,* (2004). Although it is generally believed that herbal medicines are safe and have less adverse effects, several herbal products can be toxic like alfalfa, ephedra, kava-kava etc. Regular toxicity studies should be carried out with herbal products.

In view of above, careful approach with herbal medicines and evidence-based chemical management should be followed to ensure better public health care.

Mode of action of herbal medicine

One of the most difficult issues is to establish mode of action of herbal medicine. Establishing efficacy is not sufficient in modern times. *In vitro* studies are difficult to carry out by cell culture/organ explant culture because of solubility and cytotoxicity of herbal product. In *in vivo* efficacy studies, the issue is suitable “Animal models”, what dosages should be used, etc. Pharmacokinetic studies are often difficult as we have to decide what extract, which ingredients, marker compound to use and to identify active constituents as this may be different from active compounds.

Herb to herb and herb to drugs interaction

Herbal medicines are generally considered as comparatively safer. However, many herbs exhibit side effects as well as positive and negative interactions with other drugs, herbs. Several herbs or herbal products can interact with food, allopathic medicines or pharmaceuticals. However information on this aspect is still scanty, sketchy and lack laboratory proof. Kamil (2014) has described the medicinal uses and interactive side effects of 28 herbs with allopathic medicines. Therapeutic agents have synergistic and antagonistic interactions with other pharmacological active compounds. Medicinal plants may have similar or opposite therapeutic effects eg. warfarin delays blood clot formation. Garlic, ginger, ginseng have similar effect hence care should be taken to avoid complication. St. Johns wort may affect metabolism of drugs which should be considered. There is absolutely no data on interactions of plant products with alcohol or tabaco.

Many herbs have been noted to lower sugar levels hence due care should be taken in diabetic patients. Therefore, there is urgent need to identify various positive and negative interactions with herbal products.

Increased awareness/education in the herbal medicine

General public is largely unaware of adverse health effects of herbal medicine as only positive effects are highlighted. Unfortunately the knowledge is based on observation, and there is a lack of clinical trial data. Furthermore, contradictions of herbal products are generally not mentioned which may be dangerous for public health. There is no information on correct dose, duration of treatment and sensitive population. No detailed brochure or information is provided in majority of cases. Shelf life and stability studies are lacking for some herbal products. In view of these, information on efficacy, safety and QC of herbal formulations and beneficial as well as adverse effects should be provided to consumer and general public.

Various issues like conservation of medicinal plants, genetic variability, germplasm preservation, and intellectual property rights (IPR) have not been covered in this article.

In summarizing, herbal/traditional medicines is one of the oldest system of medicine which is still in practice for centuries. This itself is a testament to real and perceived beneficial effects. In recent years, there is increased interest for use of herbal medicines due to various reasons. Medicinal plants are natural does not mean that they are not toxic and safe. This is especially valid as active ingredients are not definitely known in many cases. There is lack of information on herb/herb and herb/ drug/ food interactions. There is urgent need of pharmacovigilance in the area of herbal medicine. Last but not the least; detailed information should be provided to consumer regarding adverse effects associated with high doses/chronic use and sensitive population. Both experimental and clinical studies are needed to expand the validity and value of herbal medicine. The continuing interface between basic science, modern medicine and herbal products research will definitely provide new strategies and methodology top merge in the age of global medicine.

ACKNOWLEDGEMENT

I wish to thank all my colleagues from NITM, Belgavi especially Subarna Roy, Dr. Harsha Hegde and Dr. Satish Hegde for their help and work on medicinal plants. I wish to thank Indian Council of Medical Research for the financial support.

REFERENCES

- Hegde, S., Pai, S.R., Bhagwat, R.M., Saini, A., Rathore, P.K., Jalalpure, S.S., Hegde, H.V., Sugunan, A.P., Gupta, V.S., Kholkute, S.D., Roy, S. (2018): Genetic and phytochemical investigations for understanding population variability of the medicinally important tree *Saraca asoca* to help develop conservation strategies. *Phytochemistry*. (2018) Dec;156:43-54. doi: 10.1016/j.phytochem.2018.08.016. Epub 2018 Sep 4. PMID: 30189346.

- Hegde, S., Pai, S.R., Bhagwat, R.M., Saini, A., Rathore, P.K., Jalalpure, S.S., Hegde, H.V., Sugunan, A.P., Gupta, V.S., Kholkute, S.D., Roy, S. (2019): Population genetic and phytochemical dataset of *Saraca asoca*: A traditionally important medicinal tree. Data Brief. (2019) Jun 25;25:104173. doi: 10.1016/j.dib.2019.104173. PMID: 31516922; PMCID: PMC6728264.
- Iqbal, M. (2018). Botanical identity of traditional herbal drugs. Acdemic Staff College, JMI, New Delhi.
- Patwardhan, B. (2004). Drug, Discovery and development: Traditional medicine and ethnopharmacology. Avishkar publishers, Jaipur, India.

CHAPTER

2

Reproductive Toxicity of Herbicides

Priyanka Sharma[1], Bhawana Sharma[2], S.C. Joshi[3]*

[1] Lords University, Alwar, Rajasthan

[2] S. S. Jain Subodh PG (Autonomous) College, Jaipur, Rajasthan

[3] Reproductive Toxicology Unit, Department of Zoology, University of Rajasthan, Jaipur (Rajasthan) India.

ABSTRACT

In recent years, people have been exposed to several types of substances with broad spectrum due to the rapidly evolving technology. One of these chemical substance groups are herbicides. Herbicides have been an essential part of agriculture to protect crops and livestock from pest infestations. They are commercialized in different formulations and their residues are frequent pollutants in the environment. Despite their usefulness, herbicides could pose potential risks to food safety, the environment, and all living things. The potential for these agricultural chemicals to cause endocrine disruption (ED) in humans and wildlife is an increasing concern; however, the effects of commonly used herbicides at environmentally relevant concentrations are largely unknown. Therefore, environmentally relevant pesticides (herbicides) were tested for their endocrine disrupting potential in different animals.

A huge number of chemicals have been identified as endocrine disruptors, among several pesticides. For several years, there have been enquiries about the impact of environmental factors on the occurrence of human pathologies. Herbicides are compounds that alter the normal functioning of the endocrine system of both wildlife and humans. There has been substantial interest in the potential adverse effects of exposure to environmental hazardous chemicals, including herbicides on male reproduction and resulting declining in reproductive performance in many countries. This chapter essentially summarizes existing information on the adverse effects of herbicides on the reproductive functions in different animals and underlines the potential mechanism of reproductive toxicity induced by them.

Keywords: Endocrine disruptors; pesticides; reproductive toxicity; adverse effect

INTRODUCTION

Pesticides and a wide variety of other chemicals persist in the environment and make their way into our everyday lives. While the concept that toxicants accumulate in the body and are the cause of various health problems has long been a fundamental tenet of traditional healthcare systems around the world, researchers have learned a great deal in recent years about how toxicants affect the human population, where they originate, and how to improve our ability to detoxify in a toxic world. How can we help patients manage the health risks of exposure to these ubiquitous substances?

In the last half of the last century, worldwide pesticide production increased at a rate of about 11% per year, from 0.2 million tons in the 1950s to more than five million tons in 2000. What's more, it is estimated that only 0.1% of applied pesticides reach the target pests, leaving the vast majority of the chemicals (99.9%) to linger in the environment (El-Sayed *et al.*, 2012).

Pesticide usage has gradually increased throughout the decades; in 2017, over 4 million tonnes of pesticides were used globally. Asia (52.8%), followed by the United States (30.2 %) and Europe (13.8 %) were the continents that used the most pesticide. Insecticides, a type of insecticide, account for almost 100 million tons each year. Several studies have found a link between environmental contamination and human and animal health problems. Carcinogenesis, chronic illness, endocrine disruption, and congenital anomalies/malformations have all been linked to pesticides, plasticizers, disinfection by-products, heavy metals, and persistent organic pollutants (Katia *et al.*, 2016). Chemical and medication makers, residential sewage, polymer and petrochemical-based businesses, oil refineries, mining sites, glass blowing facilities and other anthropogenic sources of xenobiotics are only a few examples. Pesticides and heavy metals, on the other hand, are the most common xenobiotics produced by agricultural and industrial activities. To improve agriculture productivity, a number of chemicals such as insecticides, pesticides, and herbicides are utilized. Annually, 4.6 million tons of chemical pesticides are sprayed into the environment around the world (Mitra and Maitra., 2018).

By a orderly judgment of epidemiological studies concerning the group with common interest or pursuit middle from two points pesticide uncovering and mental wellness outcome, the European Food Safety Authority (EFSA) emphasize that human uncovering to pesticides is connected to a lot of mental wellness implications, containing malignant growth, neurological affliction and concerning the mind development, respiring and cardiovascular affliction, diabetes reproductive affliction and endocrine disorders (Lucia *et al.*, 2020). In particular, the concluding deserve important in own way consideration and further studies (Dziewirska *et al.*, 2018). Reproductive adverse belongings of pesticides are having connection with intercourse between animate beings. In men, pesticides putting in view or danger has happen mix to reduced testosterone combining and sperm condition and action, as well as testicular malignant growth. However, the main concern exist for children, the one are specifically open to attack to the adverse belongings of pesticides with endocrine-upset enterprise, due to their odd lifestyle, digestive practice and developmental stage, in the way that in young adulthood.

Herbicides are chemicals used to kill or damage unwanted plants or parts of them. The term derives from the Latin words herbs and caedo, meaning plant-killer (Manno., 1996). Herbicide application occurs most frequently in row-crop farming, where they are applied before or during planting to maximize crop productivity by minimizing other vegetation. They also may be applied to crops in the fall, to improve harvesting.. Herbicides are used in forest management to prepare logged areas for replanting. The total applied volume and area covered is greater but the frequency of application is much less than for farming (Shepard *et al.*, 2004). Modern herbicides can be classified by different criteria such as the chemical class, the time or site of application, the effect on the plant, the mechanism of action or toxicity. In the last decades the production and use of modern herbicides has increased faster than those of any other class of pesticides. The world production of herbicides is more than double that of all insecticides, and more than triple that of fungicides.

The potential effects of herbicides are strongly influenced by their toxic mode of action and their method of application. The molecular site of action is challenging to predict because structural associations have not been identified, but modes of action are well-established. Herbicides can act by inhibiting cell division, photosynthesis or amino acid production or by mimicking natural plant growth hormones, causing deformities (Ross and Childs., 1996).

The World Health Organization (WHO) has reported that roughly three million pesticide poisonings occur annually, resulting in 220,000 deaths worldwide. In some cases, it has been suggested that diseases such as cancer, allergies, neurological disorders and reproductive disorders may be connected to herbicide exposure (Anifandis *et al.*, 2018). In this chapter, the known acute and chronic reproductive toxicological effects of the major classes of herbicides will be reviewed.

Several environmental pollutants, including herbicides, act as endocrine disrupting chemicals (EDCs). They can cause cancer, diabetes, obesity, metabolic diseases and developmental problems (Ahmad *et al.*, 2017). The endocrine system may be the main target for the toxic manifestation of pesticides and may result in reproductive alterations, especially in steroid-hormone-dependent functions (Gourounti *et al.*, 2008). Endocrine disruptors were defined by (Kavlock *et al.*, 1996) as exogenous agents that interfere with the production, release, transport, metabolism, binding, action, or elimination of natural hormones responsible for the maintenance of homeostasis and the regulation of developmental processes. These disturbances may potentially cause risk to population by impairing their capacity to reproduce (McDaniel *et al.*, 2008), being that occupational exposure to pesticides has been linked to a reduction in the quality of semen and a greater rate of infertility and miscarriage (Weselak *et al.*, 2006).

Among the variety of toxicants that affect male fertility, fungicides, insecticides, and herbicides seem to be environmental factors that may have a negative impact on human sperm quality (Giudice., 2016). Many of these chemicals are endocrine disruptors; their bioaccumulation in humans may cause adverse health effects, especially for people living in regions in which outdoor spraying has been performed to protect crops from competing weeds. The extensive use of herbicides, given that it poses a risk to reproductive health, is a controversial matter (Sengupta and Banerjee., 2014). So that, this chapter focuses on herbicides that cause reproductive toxicity, and reviews the available information about the exposure and effects of such herbicides, as well as human biomonitoring for human health risk assessment.

The main herbicides with effects described on the male reproductive system as an endocrine disruptures are alachlor, acetochlor, atrazine, MCPA (4-chloro-2-methylphenoxy acetic acid), metolachlor, 2, 4-d, metribuzin and Paraquat. Toxic effects of these herbicides with reference to reproductive function in male experimental models are discussed below:

ALACHLOR

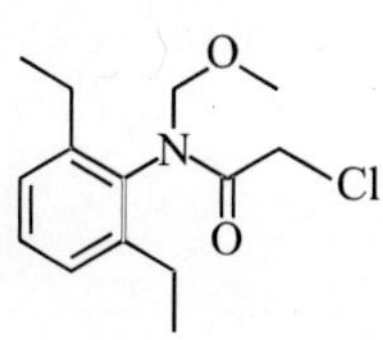

Alachlor is an acetanilide herbicide used in field corn, soybeans, and peanuts to reduce annual grasses and weeds. It's a systemic herbicide that's absorbed by sprouting shoots and roots (Miyuki *et al.*, 2003). Although the particular routes are unknown, acetanilide herbicides appear to interfere with a number of physiological processes, including lipid, protein, and flavonoid production (Kangetsu and William., 2001).

These herbicides are typically found in surface water and groundwater and are widely used in agriculture (Atmakum, and Sinthalapadi., 2004). Although alachlor has a modest acute toxicity, it has been documented that repeated exposure causes hepatotoxicity, permanent uveal degeneration, and tumour growth in some animals (Charles *et al.*, 2003).

Alachlor is also one of the herbicides that has been linked to endocrine disruption. It is a strongly suspected endocrine disruptor, as it is recognised as a probable endocrine disruptive chemical by the Environmental Protection Agency (EPA) and the World Wildlife Fund (WWF). A number of mammalian and aquatic reproductive toxicity investigations with alachlor were conducted under acute, subacute, and chronic circumstances (Chulwoo *et al.*, 2004).

ACETOCHLOR

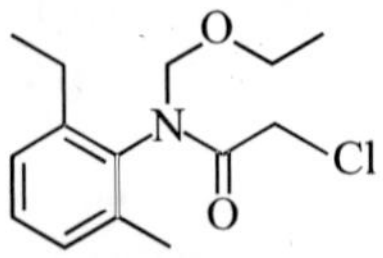

Acetochlor (2-chloro-N-(ethoxymethyl)-N-(2-ethyl-6-methylphenyl)acetamide) is a pre-emergence herbicide that belongs to the class of chloroacetyl acetanilides. It is one of the most commonly used herbicides for residential and agricultural applications (Zhao 2017). It has been extensively used all over the world and has become one of the three most frequently used herbicides in China. In addition, it is also the third most commonly applied herbicide in the agricultural market estimated by the USEPA (Wang *et al.*, 2017).

It is clear that oxidative stress-triggered unfastened radicals can function signaling molecules that cause apoptotic mobileular loss of life and inflammatory responses having an detrimental impact on steroid synthesis.eleven Therefore, this observe aimed to make clear the impact of subacute publicity to acetochlor at the male reproductive gadget in addition to the feasible interactions among oxidative stress, apoptosis and the endocrine gadget (Xianping *et al.*, 2019).

Acetochlor can induce systemic and reproductive toxicity in subacutely exposed male mice, such as clear reduction in body weight, liver function damage in mice, oxidative stress damage and histopathological degeneration in testicular tissue as the concentration of acetochlor increases (Sidorkiewicz *et al.*, 2017).

MCPA (4-CHLORO-2-METHYLPHENOXY ACETIC ACID)

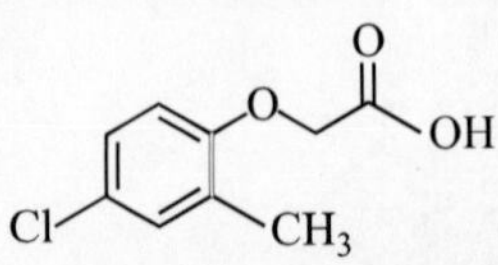

MCPA (4-Chloro-2-methylphenoxyaceti c acid), a plant growth regulator -type herbicide, which belongs to a collection of chemical substances associated with the plant increase hormone, auxin. Phenoxyalkanoic acid herbicides are compounds containing both a 2,4-dichlorophenoxy or a 4-chloro-2-methylphenoxy organization substituted onto a carbon of acetic, butyric, or propanoic acid (Paszko *et al.*, 2016).

Acute MCPA poisoning in people can reason nausea, vomiting, stomachache, diarrhea, headache, dizziness, muscle fasciculation, hypotension, dyspnea, liver and kidney dysfunction, and coma. The oral LD_{50} values of MCPA for animals variety from 300 TO 700 mg/kg, with the viable exception of dogs (50-100 mg/kg) (Tatsunori *et al.*, 2008).

METOLACHLOR

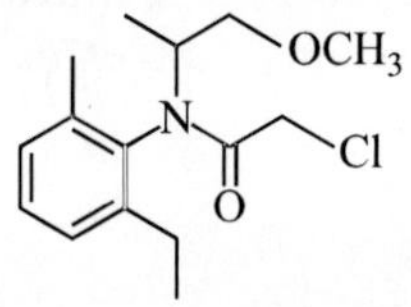

S-metolachlor [S-2-chloro-N-(2-ethyl-6-methylphenyl)-N-(2-methoxy-1-methylethyl) acetamide] is a chloro acetanilide herbicide used in maize, cotton, and soybean plantations, among other crops, to suppress weeds. Although only the S-enantiomer is biologically active, the commercial formula of S-metolachlor contains 88 % S-enantiomer and 12 % R-enantiomer. S-metolachlor (50 % S-enantiomer and 50 % R-enantiomer) has a lower danger to users, consumers, and the environment than metolachlor (50 % S-enantiomer and 50 % R-enantiomer) (Cao *et al.*, 2008 ; Liu *et al.*, 2006).

Metolachlor is metabolised and eliminated in the urine as metolachlor mercapturic acid conjugates. Its metabolite was found in a high percentage of urine samples from farm families (Curwin *et al.*, 2010) and pre-season applicators (Hines *et al.*, 2003) and linked to reproductive damage, particularly in semen quality (Swan *et al.*, 2003). Model animals have been used in a few research on the reproductive toxicity of metolachlor. Metolachlor is a widely used herbicide that might be related to effect of endocrine chemical disruptor since it was known to alter the aromatase activity in culture cells (Laville *et al.*, 2006).

(Greenlee *et al.*, 2004) discovered that in vitro incubation of embryos with metolachlor alone or in conjunction with other pesticides causes pre-implantation embryonic loss in mice, resulting in lower blastocyst development rates. The harm can occur at low doses, similar to occupational exposure and/or drinking contaminated water. (Osano *et al.*, 2002; Mai *et al.*, 2014; Varnagy *et al.*, 2003).

Exposure of metolachlor at 5 or 50 mg/kg/day for 53 days during the prepubertal period caused an increase in serum concentrations of testosterone, estradiol, and FSH and a reduction in dihydrotestosterone (DHT) but did not alter the LH. There were also observed a higher amount of fluid in the seminal vesicles, precocious puberty, and changes in morphology of the seminiferous epithelium of treated animals (Mathias *et al.*, 2012).

2, 4-D

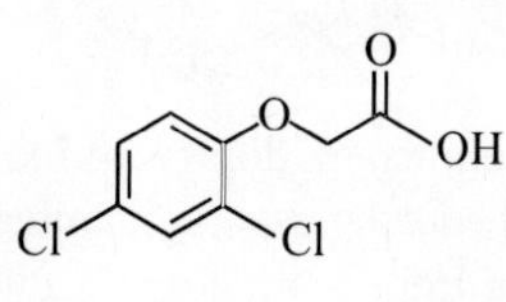

The herbicide 2,4-D is a phenoxyacetic acid-based agricultural pesticide that is commonly used to control broadleaf weeds. It is the world's most widely used herbicide. Following World War II, its popularity skyrocketed. Humans and animals are exposed to it via drinking water, soil, air, the food chain and during production of the herbicide (Carvalho., 2017).

The residue on agricultural products causes serious damage in humans and animals ingesting these items, as well as in the entire ecosystem, when 2,4-D is used irresponsibly and inappropriately. According to studies in the field of reproductive toxicology, the herbicide affected spermatogenesis and produced structural alterations in male reproductive organs, resulting in a drop in sperm count (Vural *et al.*, 2016).

METRIBUZIN

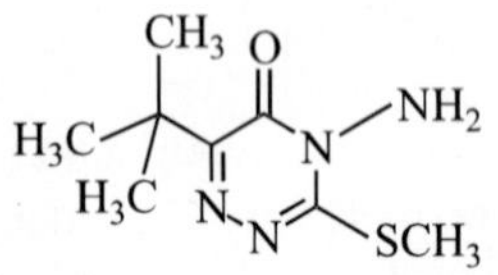

Metribuzin is a synthetic organic molecule used as a selective triazinone herbicide in wheat, potato, tomato, and other crops as a post-emergent soil applied herbicide. The goal of this research was to see how metribuzin affected the environment and human health. Humans are exposed to metribuzin by inhalation and ingestion, which occurs most commonly in agricultural settings.

In adult rats, quercetin reduces the harmful effect of metribuzin-enhanced haematological and biochemical effects. Metribuzin also caused oxidative stress in tissues by increasing malondialdehyde levels, lowering glutathione peroxidase activity, and lowering reduced glutathione levels (Kadeche *et al.*, 2016).

According to (Koutnik *et al.*, 2014) metribuzin increased oxidative stress measures such as TBRS (thiobarbituric acid reactive substances), whereas the levels of SOD and catalase were considerably reduced in a 30-day research in cray fish at a dose level of 3.06 mg/l. Sencor (a herbicide containing metribuzin) increases SOD activity while decreasing catalase, GPx, and GR activity in the liver and kidney.

The subchronic toxicity of metribuzine on common carp developmental stages and larvae was studied for 30 days at concentrations of 0.9, 4, 14, and 32 mg/L, and high concentrations resulted in increased mortality as well as histological abnormalities in the caudal kidney and liver (Stepanov *et al.*, 2012).

PENDIMETHALIN

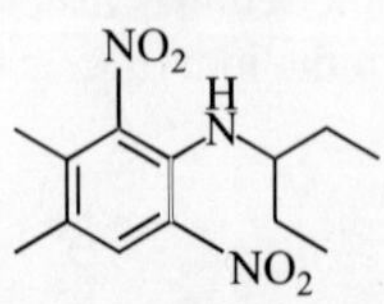

Pendimethalin (PND) is a member of dinitroaniline herbicide. It is widely used to control crop and non-crop areas to for landscape maintenance and lawn care. Herbicides are extensively used chemicals for improving crop production in modern day agriculture. India is the 4th largest pesticide producing country in the world and ranked 2nd in the Asia.

Estimated annual consumption of PND in 2014 was ~10 million pounds (Ahmad *et al.*, 2016).

PND has been detected as a contaminant in the water sources in the Denmark, France, Spain and United. (Ahmad *et al.*, 2017) also reported the endocrine disrupting potential of pendimethalin with its probable antiandrogenic function via in silico study. PND induces cellular toxicity and genetic perturbations which can alter the normal cellular and physiological functioning of testes in rats.

Atrazine

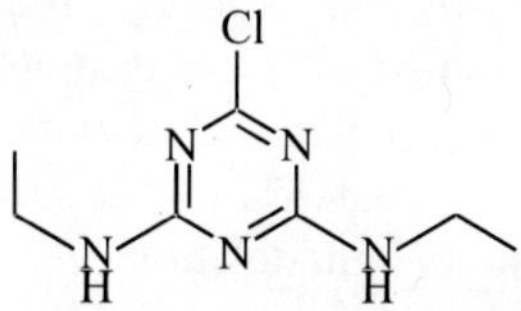

Atrazine (2-chloro-4-ethylamino-6-isoprop- ylamino-s-triazine, ATZ), a chloro-s-trazine herbicide, is used extensively worldwide for broadleaf and grassy weed control in corn, sorghum, sugarcane, cotton, and pineapple crops and landscape vegetation (Swan, 2006), so that ATZ and its metabolites are widely persistent in water and are mostly found in soil especially in farming seasons. Despite ATZ has been banned in European Union and been restricted in other countries, it is still being used in large quantities worldwide up to now. It is one of the most widely used agricultural pesticides in United States, and its application in Asian countries has been growing. Therefore, humans and wildlife are at risk for exposure to ATZ (Barr *et al.*, 2007).

It was reported that ATZ exposure from postnatal day (pnd) 21 to 53 had caused a significant decrease in serum and testicular T levels after administered at doses of 100-200 mg/kg bw/day in male Wistar rats (Stoker *et al.*, 2000a). A recent study demonstrated that the serum levels of testosterone (T), follicle stimulating hormone (FSH), luteinizing hormone (LH), and inhibin-B (INH-B) had decreased by 85% after 48 d exposure to high dose (300 mg/kg bw/day) of ATZ in adult male Wistar rats (Dehkhargani *et al.*, 2012).

These results indicates that relatively high dose range of ATZ could damage the seminiferous tubule; influence the production of spermatozoa; decrease the activity of testicular marker enzymes; induce oxidative stress; and disturb the secretions of T, FSH, LH, and INH-B. It could be concluded that ATZ of relatively high dose range could induce reproductive toxicity in male rats (Yang *et al.*, 2014).

Paraquat

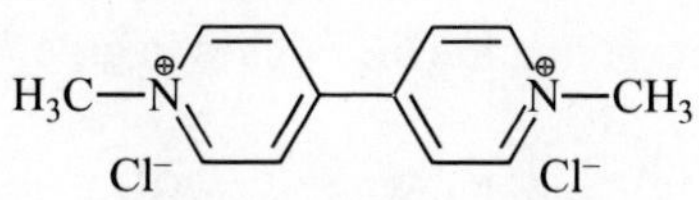

Paraquat (PQ) is a herbicide which is used all around the world. It has chemical formula as N_1, N paraquat-dimethyl 4, 4 dipyridine that is very toxic to humans and animals with a delay mechanism that this mechanism produce anion super oxide. PQ is consumed and produced very much in the world because it has unique characteristics and is a very powerful herbicide (Hamay *et al.*, 2012).

It was shown that PQ can caused a decrease in the weight of the male reproductive organs in rat such as testes, epididymis, prostate and seminal vesicle and the number of sperms, cell and also decrease in spermatogonia cells (Zain., 2007). Studies showed that PQ causes decrease in the quantity of sexual cells among Bul/B Syrian rats and also it can wrinkle sexual cells (Ranjbar *et al.*, 2014). High dose of PQ have significant effect on spermatocyt, spermatid, leydig cells and parameters of seminiferous tubules because of free radicals present in PQ, which have toxic and significant effects on number of spermatogenic cells and parameters of seminiferous tubules (Fathi *et al.*, 2015).

(Parisotto *et al.*, 2018) showed that the number, viability and total motility of sperm cells were significantly decreased by PQ exposure. Indeed, we observed an abnormal morphology; identifying teratospermia in PQ exposed animals. They also observed GSH depletion, which was associated with decreased activity of SOD and MPO, while CAT and GST were unaltered by PQ treatment. The PQ-induced male reproductive toxicity is associated with oxidative stress and p53 modulation.

CONCLUSION

This chapter indicate that the exposure of herbicides induced abnormalities in reproductive system which may be as a result of disturbance in the androgen-estrogen balance, as well as oxidative stress and impairment in testicular functions. All these events lead to a perceptible change in sex hormone balance and spermatogenesis arrest. Herbicide exposure affects spermatogenesis leading to poor semen quality and reduced male fertility but these effects are dependent on dose and duration. Thus, application of herbicides should be limited to a designed program with special care in handling to limit or minimize hazards to both wild life and humans.

REFERENCES

- El-Sayed, M.A., El-Damaty, A., Abdel-Razik, H. F., Gamal, R., Hany, M. F. (2012): Biochemical and Histopathological Effects of Systemic Pesticides on Some Functional Organs of Male Albino Rats. *J Appl Sci Res,* 8(11): 5459-5469.
- Katia, C.D.M., Tavares, V., Jennifer, C.C., Edson, Z., Jose, M., Sanches, J. and Ana, P. A. F. (2016): Maternal and fetal toxicity of wistar rats exposed to herbicide metolachlor. Acta Scientiarum Biological Sciences, 38(1)91-98.
- Mitra, A. and Maitra, S. K. (2018): Reproductive Toxicity of Organophosphate Pesticides. *Annals of Clinical Toxicology.* 1 (1): 1-8.
- Lucia, C., Sabrina, T., Lorella, C., Gianluca, F., Carmine. M., Monia, P., Enrica, F. and Cinzia, L.R. (2020): Integrated approach to evaluate the association between exposure to pesticides and idiopathic premature the larches in girls: The Peach Project. *Int J Mol Sci,* 21: 1-12.
- Dziewirska, E., Hanke, W., and Jurewicz, J. (2018): Environmental non-persistent endocrine-disrupting chemicals exposure and reproductive hormones levels in adult men. *Int J Occup Med Environ Health,* 31: 551-573.
- Manno, M. (1996). Herbicides. *Human Toxicology,* 551-560.
- Shepard, J.P., Creighton, J. and Duzan, H. (2004): Forestry herbicides in the United States: an overview. *Wildlife Society Bulletin* 32(4):1020-1027.

- Ross, M. A. and Childs, D. J. (1996): Herbicide mode-of-action summary. Purdue University, Department of Botany: Plant Pathology, West Lafayette IN. Report No. W-5-W-23.
- WHO (1992). Our Planet, Our Health; Report of the WHO Commission on Health and Environment; WHO: Geneva, Switzerland.
- Anifandis, G., Amiridis, G., Dafopoulos, K., Daponte, A., Dovolou E., Gavriil, E., Gorgogietas, V., Kachpani, E., Mamuris, Z., Messini, C. I., Vassiou, K. and Psarra, A. M. G. (2018). The in vitro impact of the herbicide roundup on human sperm motility and sperm mitochondria. *Toxics,* 6(1): 2.
- Ahmad, I., Usman, A., and Ahmad, M., (2017). Computational study involving identification of endocrine disrupting potential of herbicides: Its implication in TDS and cancer progression in CRPC patients. *Chemosphere,* 173: 395-403.
- Gourounti, K., Lykeridou, K., Protopapa, E., Lazaris, A., (2008). Mechanisms of actions of organochlorine substances. *J Health sci,* 2(2): 89-98.
- Kavlock, R.J., Daston, G.P., and DeRosa, C. (1996). Research needs for the risk assessment of health and environmental effects of endocrine disrupters: a report of the U.S. EPA-sponsored workshop. *Environmental Health Perspectives,* 104(4): 715-740.
- McDaniel, T. V., Martin, P.A. and Struger, J. (2008). Potential endocrine disruption of sexual development in free ranging male northern leopard frogs (Rana pipiens) and green frogs (Rana clamitans) from areas of intensive row crop agriculture. *Aquatic Toxicology,* 88(4): 230–242.
- Weselak, M., Arbuckle, T.E., and Foster, W. (2006): Pesticide exposures and developmental outcomes: the epidemiological evidence. *Journal of Toxicology and Environmental Health B,* 10(1-2): 41-80.
- Giudice, L. C. (2016). Environmental toxicants: Hidden players on the reproductive stage. *Fertil Steril.* 106: 791-794.
- Sengupta, P. and Banerjee, R. (2014): Environmental toxins: Alarming impacts of pesticides on male fertility. *Hum Exp Toxicol,* 33: 1017-1039.
- Miyuki, C., Yasuhiko, H. and Eiichi, T. (2003): Environmetal Tox And Pharm, 152: 159-166.
- Kangetsu, H. and William, T.M., (2001): Pesticide Biochemistry and Physiology, 71: 116-123.
- Atmakum, R. and Sinthalapadi, T.M. (2004): Chemosphere, 54: 647-652.
- Charles, W., David, W.G., Gina, B., Frank, D., Ben, J.C. and Cynthia, K.L. (2003): Water Research, 37: 4761-4769.
- Chulwoo, L., Jisung, R., Soo-Young, P., Kyunghee, C., Sung-Hwan, J., Jin Gyun, N. and Deok-Gil, R. (2004): Effects of alachlor on the early development and induction of estroen-reponsive genes in medaka, Oryzias latipes. *Developmental and Reproductive Toxicity,* 66: 2951-2957.
- Zhao, P., (2017): Global pesticide market overview and development trend in 2015. *Pesticide,* 02: 79-85.
- Wang, J, Li, Y., Wang, X. and Wang, Y (2017): The current situation and prospects of pesticide use in China. *Outlook Agric,* 02: 56-60.

- Xianping, S., Feng, Z., Dongya, C., Qian B., Hengdong, Z., Xin, L. and Baoli, Z. (2019): Study on systemic and reproductive toxicity of acetochlor in male mice. *Toxicology Research,* 8(1): 77-89.
- Sidorkiewicz, K., Zareba, S., Wolczynski, K. and Czerniecki, J. (2017): Endocrine-disrupting chemicals-mechanisms of action on male reproductive system. *Toxicol Ind Health,* 33: 601-609.
- Paszko, T., Muszynski, P., Materska, M., Bojanowska, M., Kostecka, M. and Jackowska I (2016): Adsorption and degradation of phenoxyalkanoic acid herbicides in soils: A review. *Environmental Toxicology and Chemistry,* 35(2): 271-286.
- Tatsunori, T., Takahito, H., Yuko, I., Mizuho, N., Shinya, M., Tomoko, M., Mariko, K., Akihiko, K. and Toshikazu, K. (2008): A Fatal Intoxication from Ingestion of 2-Methyl-4-Chlorophenoxyacetic Acid (MCPA). *Journal Analytical Toxicology,* 32: 187-191.
- Cao, P., Wang, X., Liu, F., Zhao, E. and Han, L. (2008): Dissipation and residue of S- metolachlor in maize and soil. *Bulletin of Environmental Contamination and Toxicology,* 80(5): 391-394.
- Liu, H, Ye, W., Zhan, X., and Liu, W. (2006): A comparative study of rac- and S-metolachlor toxicity to Daphnia magna. *Ecotoxicology and Environmental Safety,* 63(3): 451-455.
- Curwin, B. D., Hein, M. J., Barr D B and Striley C (2010): Comparison of immunoassay and HPLCMS/ MS used to measure urinary metabolites of atrazine, metolachlor, and chlorpyrifos from farmers and non-farmers in Iowa. *Journal of Exposure Science and Environmental Epidemiology,* 20(2): 205-212.
- Hines, C.J., Deddens, J.A., Striley, C.A.F., Biagini, R. E., Shoemaker, D. A., Brown, K. K. and Hull, R.D. (2003): Biological monitoring for selected herbicide biomarkers in the urine of exposed custom applicators: application of mixed-effect models. *The Annals of Occupational Hygiene,* 47(6): 503-517.
- Swan, S. H., Kruse, R. L., Liu, F. Barr, D. B., Drobnis, E.Z. and Redmon, J. B. (2003): Overstreet, Study For Future Families Research Group. Semen quality in relation to biomarkers of pesticide exposure. *Environmental Health Perspective,* 111(12): 1478-1484.
- Laville, N., Balaguer, P. and Brion, F. (2006): Modulation of aromatase activity and mRNA by various selected pesticides in the human choriocarcinoma JEG-3 cell line. *Toxicology,* 228(1): 98-108.
- Greenlee, A. R., Ellis, T. M. and Berg, R. L. (2004): Low dose agrochemicals and lawn-care pesticides induce developmental toxicity in murine pre implantation embryos. *Environmental Health Perspectives,* 112(6): 703-709.
- Osano, O., Admiraal, W. and Otieno, D. (2002): Developmental disorders in embryos of the frog *Xenopus laevis* induced by chloroacetanilide herbicides and their degradation products. *Environmental Toxicology and Chemistry,* 21(2): 375-379.
- Mai, H., Gonzalez, P., Pardon, P., Tapie, N., Budzinski, H., Cachot, J. and Morin, B. (2014): Comparative responses of sperm cells and embryos of Pacific oyster (Crassostrea gigas) to exposure to metolachlor and its degradation products. *Aquatic Toxicology,* 147: 48-56.

- Várnagy, L., Budai, P., Fejes, S., Susan, M., Fancsi, T., Keseru, M. and Szabo, R. (2003): Toxicity and degradation of metolachlor (Dual Gold 960 EC) in chicken embryos. *Communication in Agricultural and Applied Biological Sciences,* 68(4): 807-811.
- Mathias, F. T., Romano, R. M., Sleiman, H. K., Oliveira, C. A. D. and Romano, M. A. (2012): Herbicide metolachlor causes changes in reproductive endocrinology of male wistar rats. ISRN *Toxicology* Article ID 130846: 1- 7.
- Carvalho, F. P. (2017): Pesticides, environment and food safety. *Food Energy Secur,* 2 (6): 48-60.
- Vural, H. A., Baran, A. and Altiner, A. (2016): General and Reproductive Toxicity of 2, 4-Dichlorophenoxyacetic acid (2, 4-D). *Advances in Clinical Toxicology,* 1-4.
- Kadeche, L., Bourogaa, E., Boumendjel, A. Djeffal, A. Abdennour, C., Feki, A. E.l. and Messarah, M. (2016): Quercetin attenuates metribuzin-induced biochemical and hematological toxicity in adult rats. *Int J Pharm Sci Rev Res,* 40(1): 38-46.
- Koutnik, D., Stara, A., Zuskova, E., Kouba, A. and Velisek, J. (2014): The effect of subchronic metribuzin exposure to signal crayfish (*Pacifastacus leniusculus* Dana 1852). *Neuroendocrinology Letters,* 35: 2.
- Stepanov, S., Dolezelov, P., Plhalov, L., Prokes, M., Marsalek, P., Skoric, M., Svobodov, Z. (2012): The effects of metribuzin on early life stages of common carp (Cyprinus carpio). *Pesticide Biochemistry and Physiology,* 103: 152-158 .
- Ahmad, I., Ahmad, A. and Ahmad, M. (2016): Binding properties of pendimethalin herbicide to DNA: multispectroscopic and molecular docking approaches. *Phys Chem Chem Phys,* 18: 6476-6485.
- Ahmad, M. I., Usman, A. and Ahmad, M. (2017): Computational study involving identification of endocrine disrupting potential of herbicides: Its implication in TDS and cancer progression in CRPC patients. *Chemos,* 173: 395-403.
- Swan, S. H. (2006): Semen quality in ferlile US men in relation to geographical area and pesticide exposure. *Int J Androl,* 29: 62-68.
- Barr, D. B., Panuwet, P. and Nguyen, J. V. (2007): Assessing Exposure to atrazine and its metabolites using biomonitoring. *Environ Health Persp,* 115: 1474-1478.
- Stoker, T. E., Laws, S. C. and Guidici, D. L. (2000a): The effects of atrazine on puberty in male Wistar rats: An evaluation in the protocol for the assessment of pubertal development and thyroid function. *Toxicol Sci,* 2000a, 58: 50-59
- Dehkhargani, S. F., Shahrooz R and Malekinejad H (2012): Atrazine in sub-acute exposure results in sperm DNA disintegrity and nuclear immaturity in rats. *Veterinary Research Forum,* 3: 19-26.
- Song, Y., Jia, Z.C., Hu, J, X., and Zhaf, L.S.Z.C., (2014) Toxic Effects of Atrazine on Reproductive System of Male Rats. *Biomed Environ Sci,* 27(4): 281-288.
- Hamay, A.V., Najafiyan, M., Farzam, M. and Kargar, H. (2012): Effects of the paraquat herbicide on oogenesis and ovary structure of wistar rat strain. *Advances in Environmental Biology,* 6(3): 1006-1012.
- Zain, A. (2007). The evaluation of the toxic effect of paraquat and its mechanism of action on reproductive system of male rats: Thesis. Universiti Sains Malaysia, Penang. http://eprints. usm..

- Ranjbar, A., Mohsenzadeh, F., Chehregani, A., Khajavi, F., Zijoud, S., Ghasemi H (2014): Ameliorative effect of Matricaria chamomilla. L on paraquat: Induced oxidative damage in lung rats. *Pharmacognosy Research,* 6(3): 199.
- Fathi, N., Hossinipanah, M., Hajihossini, M. and Ranjbar, A (2015): Effects of paraquat on testicular histomorphometry of male rats. Biological Forum- *An International Journal,* 7(2): 573-575.
- Parisotto, E. B., MartinsSilva, N. G., Wajima, C. S., Rahn, B., Domingues, J. T. H., Cesconetto, P. A., Cattani, D., Rieg, C. E. H., Hellbrugge, A., Karina, B. F., Rozangela, C. P., Ariane, Z. (2018): Effects of paraquat in the reproductive function and on redox state of adult male rats. *Free Radical Biology and Medicine,* 120(1): S88-S89.

CHAPTER 3

Studies on the Impact of Construction of Barrage on the Water, Sediment and Biota of Central Himalayan River, Tawi in the Vicinity of Jammu City

Sharma K.K[1], Shvetambri[2*] and Neha Sharma[3]
**Email: shvetambrijasrotia@gmail.com*
1. University of Jammu
2. Central University of Jammu
3. University of Jammu

ABSTRACT

The natural resources of fresh water such as rivers hold immense importance as they play a critical role in shaping the existence of life on Earth. Since times immemorial civilizations have settled around the rivers. Human ancestors stored/distributed the water by making crude structures in the form of impoundments, pools, drains, gates, bunds, dams, sluice etc. With the progressions in human intellect the technology became more advanced and water was made available in plenty to meet the needs of domestic, agricultural, industrial and hydroelectric sectors. The demand for clean water has infinitely increased in the past centuries with exponential increase in the population. In the present-day structures such as dams, barrages, bundhs etc. which were earlier built with an aim to preserve and conserve water has become controversial leading to conflict between social groups and disputes between different knowledge systems. It is believed that dams/barrages/hydro-technological systems built on rivers have fallen prey to the multinational donor agencies, global policy shifts, international/national human rights courts, local and international civil society movements, indigenous groups, environmental Non-Government Organizations, media, government and bureaucratic agencies, experts, scholars and activists. Therefore, it becomes imperative to check the impact of these structures on the health of the rivers by regular assessments/studies. A similar study was carried out to study the impact of Barrage construction on water, sediment and biota of River Tawi. The study revealed that construction of barrage created stress on the ecology of the river. It was recorded that the river was at the receiving end of several habitat destructive activities including dredging, channel modification, exposure and removal of sediments. These disturbances further resulted in altered channel width to depth ratios, decreased diversity and abundance of biota, decreased dissolved oxygen concentration with increasing water temperatures and overall reduction in water quality and loss of floodplain connectivity. The zooplankton, macrobenthic invertebrates and fish fauna recorded from the river during the two-year study period was observed to be hardy with an ability to survive in wide range of flow, pH, and temperature fluctuations along with having elaborate reproductive strategies (vegetative propagation), invasive and pollution tolerant thereby indicating the polluted and disturbed status of the river.

Keywords: Rivers, Barrage, Impact Assessment studies.

INTRODUCTION

Rivers are the key component of human societies from earlier days of civilization and contributed to agriculture as well as public resources. Apart from supporting water for drinking, irrigation, transportation, and hydroelectricity rivers have great ecological significance as they not only provide habitat to a huge diversity of aquatic biota, but also play a role in nutrient and sediment transport. There cultural, spiritual, and recreational value cannot be denied. Riverine systems are unarguably one of the most precious resources on earth and yet most poorly managed in the wake of increasing demand of freshwater due to population explosion. According to WWF Report (2018), demand for freshwater has been increasing at a rate of 1% per year and will continue to increase due to shifting consumption patterns.

Flowing water ecosystems face numerous hurdles, from increase in direct utilization and contamination due to unceasingly cumulative population and rise in living standards, to both direct and indirect disturbances to river column and bed by river regulation, flow modifications, building of dams, barrages, fragmentation, channelization etc. According to Millennium Ecosystem Assessment 2005, all these activities have resulted in massive loss of biodiversity since 1970. Modifications of river channel, flows and flood plains of rivers for various purposes poses immense short term as well as long term threats to riverine ecosystems. Rivers are more than just conduits of water as they sustain the aquatic as well as amphibious biota, transform geo-morphological landscapes as they move and play an important role in nutrient cycling (Opperman *et al.* 2018). So any perturbation or disturbance to the essential structure of rivers also affects its wide array of natural functions.

Whenever there is a distinct and abnormal change to the properties of an ecosystem due to any kind of disturbance, it is termed as perturbation (Bender *et al.* 1984). Disturbances can be observed as a changed in the biotic response and abiotic parameters, and can be differentiated on the basis of duration and level of intensity. (Bender *et al.* 1984; Lake 2000) A disturbance may be a discrete event or a cascading change in one of more controlling factors that eventually exceeds a critical threshold. (Sparks *et al.* 1990). As reported by Bunn and Arthington (2002), alteration of natural flow regime of a river can affect the habitat composition, disruption of life history patterns of species, loss of biodiversity by loss of native species and proliferation of invasive species along with a shift in the physicochemical parameters of water and sediments, thereby further affecting biotic responses and adaptations. Disturbances as a result of river channelization and channel modification have also been reported to disrupt stream equilibrium, riffle and pool habitat, change river velocities, eliminate the function of floods to control channel-forming properties, increase erosion and sediment load and alter the river bed base level (US EPA, 2007).

The pace and scale of hydrological alterations have not only impaired the riverine ecosystems globally but also questioned the ability of ecologists to understand and mitigate such impacts (Poff *et al.* 2010). Therefore, careful and comprehensive impact assessment studies need to be conducted to understand the intensity and extent of such disturbances.

In Jammu city, a similar project was undertaken on river Tawi, where construction of a barrage on the river was proposed to partially convert the river to a lentic system or an artificial lake. The project which was initially a part of enhancement in tourism sector by the then state Govt. started in late 2008 and picked up pace during 2012-14. Although no permanent impoundment was created but significant water diversions, channelization, riverbed dredging, river bank development activities along with construction activity for the erection of pillars for the barrage in the river bed exerted a significant impact on the already distressed ecosystem of river Tawi. Since the river already faces large pollution load from the city itself along with the land use patterns along the banks, studying the impact of this further disturbance on the river becomes more vital.

Study Area

The River Tawi also called as "Surya Putri" is an important topographic feature of the state, originates from the Central Himalayan axis (Dhauladhar range) at an elevation of about 1400 feet. The point of origin lies below Seoj Dhar peak, from the lap of Kali Kundi glacier and the adjoining area of Southwest of Bhaderwah in Doda district. It has latitudinal position 32^0 35' - 33^0 5' N and longitudinal position 74^0 35' - 75 45' E and it is the left bank tributary of river Chenab. It flows between Jug dhar and Trishul dhar in a westerly direction till Udhampur where it takes a southerly bend across the Shivalik range and again resumes a westerly course passing along the Jammu city and it joins Chenab River at Marala in POK. It is fed by a number of tributaries during its course. The total catchment area of the river Tawi is more than 2,100 square kilometers. Total stretch of the river adds up to 141 kilometers and traverses through Doda, Udhampur and Jammu.

The flow of river is considerable and variable throughout the year. River is very prone of flash floods especially at the time of monsoon rains and also at the time of irregular winter rains. Carries a large mass silt during rainy season. Topography of river is rugged and is surrounded by moderate to thinly forested slopes.

River Tawi is one of the most important sources of drinking water for the people of Jammu city as about 26 mgd (million gallons per day) water from three water treatment plants Sitlee (20 mgd), Dhountli (4 mgd) and Boria (2 mgd) is pumped from the river to cater to the needs of drinking water in Jammu. Along with drinking uses, river water is also used for irrigation, fishing and recreation etc. River is under pressure owing to anthropogenic stress at the city bridge sites, untreated sewage disposal from the nearby colonies, garbage, animal excreta, dead animals, agricultural runoff, detergents, addition of pollutants from its tributaries along with pollution by various religious activities.

As reported by Jammu and Kashmir State Pollution Control Board (JKSPCB), nearly 18 sewage channels are the direct source of non-biodegradable wastes such as polythene and organic sewage into the Tawi. Solid waste disposal at or near the banks and in the river, religious waste, bathing of cattle, washing clothes are also the sources of pollution for River Tawi. In addition, waste from the Gujjar Nagar (Jogi Gate crematorium) and encroachment on the banks further worsen the pollution problem. Frequent dredging of the bottom and shores for extraction of sand and stones has altered shoreline morphology over the years. The water level of the river has already declined to great extent with most of the river looking like a dry patch during the summers.

As a part of water infrastructure development scheme of J&K state Govt. Under the "Artificial Lake Project" of river, pondage of 14,13,000 cum (0.0011 MAF) was under construction across the river, by obstructing the flow of Tawi by way of auto mechanically cum manually operated fully gated barrage of 4 meters height. The creation of artificial lake or pondage of about 1570 meters length and 450 meters average width on River Tawi was started as an integrated tourism project and augmentation of the Ranbir irrigation canal (source of water River Chenab) during the dry period. It was proposed to serve as a source of attraction for birds, also thought to help in bringing down the temperature of the city to some extent and also help in recharging ground waters. Along with these agricultural and ecological roles, the project was proposed to provide the recreational facilities like fishing, boating, water sports etc.

The construction of barrage having a length of 370 meters was undertaken in lower reaches of river Tawi. Barrage was divided into two parts since an island or land mass separates river Tawi into two distributaries, i.e., Waddi Tawi and Nikki Tawi, which were named in the local language according to the water mass they carry throughout the year. Waddi Tawi being narrower but carried larger water mass while Nikki Tawi being wider but carried comparatively lesser water mass. First stretch of 137 meters of barrage pillars were constructed across Waddi Tawi between right bank of the river upto the island and the second stretch of 237 meters across Nikki Tawi from the island towards the left bank of the river.

More than 90% of construction work of left and right embankments for the barrage, raising piers and developing the bed by way of constructing concrete blocks and concrete cistern with spillway and gangway along the alignment of the barrage was complete. Bunds were raised across the river to prevent water from entering into the area where work is in progress. Along with this, various kinds of machinery, vehicles etc. were continuously at work on the barrage site. Due to temporary impoundments at the construction site, velocity of the river in that particular stretch was considerably reduced leading to increased sedimentation and reduction in depth.

A study was designed to study the impact of anthropogenic stress on the river in view of ongoing construction of barrage on the river during a time period of two years 2013-15. Four stations were selected on the river, two upstream, one at the barrage site and one downstream the barrage. All the stations were selected in the vicinity of Jammu city as impacts on the river from all the combined anthropogenic factors are maximum at this particular stretch of river. Pollution was the common denominator at all four stations, with organic and inorganic pollution from various sources being the derogatory factor at upstream stations while downstream stations being at the receiving end of all the disturbances to the river bed and riparian areas due to construction of barrage along with direct and indirect influx of pollutants.

Station-I

Situated at Gujjar Nagar, this station receives effluents from Jogi gate crematorium, butcher-shops from Gujjar Nagar and sewage canals from the adjoining residential areas. Along with these, the station is a dumping ground for garbage, clay heaps from road construction areas near the banks and also has several bank development activities such as construction of concrete or levee banks etc. and culverts near the banks.

Station-II

It is situated near Sanjay Nagar Mosque about 855 meters away from St-I. This station receives organic sewage and garbage from houses at the encroached bank and receives maximum religious pollution load too. This station does not have any river channel modifying activity therefore sole source of perturbation at this station is pollution.

Station-III

This station is situated on the barrage-construction site, situated about 1.4 Kilometers from St-II. In addition to pollution from sewage canals or drains from the city which enter this station directly at various points and dumping of solid waste, this station receives maximum physical disturbance to the river bed owing to the barrage construction, water diversions and changes in flow regimes of water.

Station-IV

Situated downstream the barrage is Bhagwati Nagar Station, about 760 metres from St-III. At this station, river Tawi shows fragmentation into small pools and muddy or marshy areas especially during low flows as a result of water diversions and impoundments at St-III. In addition, this station also receives organic sewage and garbage from the housings along the banks, run-off from agricultural and animal shelters and dairies near the banks, encroachment and dumping of solid construction waste on the banks as well as the river channel. Accumulation of organic and inorganic waste due to reduced water velocity as a result of various bunds aggravates the pollution problem at this station.

METHODOLOGY

I. Abiotic Parameters

(a) Water
(b) Sediments

(a) Water:

Sampling and Analysis: Water samples were collected on the monthly basis in glass bottles from each station. Estimation of some of the physico-chemical parameters viz; water temperature, depth, transparency, velocity, pH, DO, FCO_2, CO_3^{2-}, HCO_3^- were done on the spot whereas analysis of Cl^{1-}, Ca^{2+}, Mg^{2+}, NO_3^{1-}, PO_4^{3-}, SO_4^{2-} and BOD of water samples were carried out in laboratory.

The various methods used in physico-chemical analysis of water are as under:

(i) Air and Water temperature: These were recorded with the help of a mercury bulb thermometer. For recording air temperature, direct exposure of thermometer to sunlight was avoided and it was kept in shade whereas for recording water temperature, the knob of thermometer was immersed vertically into the water for a minimum of 1 minute (Welch, 1952).

(ii) Depth: It was measured by using a graduated meter rod (Adoni, 1985).

(iii) Transparency: It was determined by using Secchi disc of 20cm diameter (painted black and white alternately on its upper surface) and a rope tied at its centre. It was computed by the following formula (Welch, 1952):

(iv) Velocity: It was measured by plastic ball method.

(v) **pH:** It was determined with the help of a portable digital pH meter (Hanna).

(vi) **Dissolved Oxygen (DO):** It was determined by Sodium Azide Modification of Winkler's method (APHA, 1985).

(vii) **Free Carbon Dioxide (FCO_2):** It was determined by using titrimetric method (APHA, 1985).

(viii) **Carbonates and Bicarbonates:** These were estimated by following APHA, 1985.

(ix) **Chlorides:** Chloride analysis was done by employing Argentometric method (APHA, 1985).

(x) **Calcium and Magnesium:** Ca^{2+} and Mg^{2+} estimation was done by using EDTA-titrimetric method (APHA, 1985).

(xi) **BOD:** It was estimated by following APHA, 1985.

(xii) **Nitrates:** Nitrates were determined by Phenol disulphonic method using spectrophotometer (APHA, 1985; Adoni, 1985).

(xiii) **Phosphates:** Phosphate content was determined by Stannous Chloride method using spectrophotometer (APHA, 1985; Adoni, 1985).

(xiv) **Sulphates:** Sulphates were determined by Turbiditimetric method using spectrophotometer (APHA, 1985; Adoni, 1985).

(b) Sediments

Sampling and Analysis: Sediments were collected from the shore of the river using a sampler once in a month from each sampling station during the two year study period and were stored in well labelled zip lock polyethylene bags and kept in an ice chest box before transferring to the laboratory. Samples were air dried at room temperature in the laboratory. Some of the parameters were assessed using air dried sediments but for remaining parameters, samples were dried in the oven at 60 ^{O}C and finally stored in air tight containers for further analysis. The chemical analysis of sediments for some of the parameters was carried out using following methods:

(i) **Sediment temperature:** These were recorded with the help of a mercury bulb thermometer.

(ii) **Sediment pH:** It was determined with the help of a portable digital pH meter (Hanna).

(iii) **Bicarbonate:** These were estimated following standard methods by APHA, 1985 and Adoni, 1985.

(iv) **Chloride:** It was analyzed by following Argentometric method (Chopra and Kanwar, 1967).

(v) **Calcium and Magnesium:** These were estimated by following EDTA- Titrimetric method (Chopra and Kanwar, 1967).

(vi) **Nitrate:** Nitrate content of the sediment samples was analyzed following Adoni (1985).

(vii) **Phosphate:** Phosphate was extracted using Sulphuric acid and determined spectrophotometrically (Adoni, 1985).

(viii) **Sulphate:** It was estimated by extracting the sediments with calcium chloride (Williams and Steinberg, 1959) and measured using turbidimetric method.

II. Biotic Parameters

(a) Plankton
(b) Macro-benthic invertebrates
(c) Fishes
(d) Macrophytes

(a) Plankton

Both phytoplankton and zooplankton were collected by filtering 50 liters of water through plankton net (Nytex 70 micro metre mesh size). The filtrate was transferred to glass vials and was preserved in 5% formalin. Qualitative analysis was done by following Edmondson and Winberg (1971); Pennak (1978) and Adoni (1985). Quantitative analysis was done by drop count method.

(b) Macro-benthic invertebrates

For the collection of macro-benthos, bottom collections from four stations were made using a sampler and the collected sample was washed through sieve no. 40 (256 meshes/ cm^2). The washed sediment samples with macro-benthos were poured into a white enamel tray and sorted in the laboratory. For effective sorting, moderate volume of water was added into the container to improve visibility. Forceps were used to pick larger organisms whereas smaller ones were sorted out using soft brush. Organisms thus obtained were then transferred into wide mouth labelled plastic containers and preserved in 5% formalin. The preserved benthos were later identified under light and stereo dissecting microscope and counted. Qualitative analysis of macrobenthos was done by using the keys by Ward and Whipple (1959), Pennak (1978), Tonapi (1980), Adoni (1985). Quantitative analysis of macrobenthos was done by using following formula:

$$O = \frac{N}{A \times S} \times 1000$$

Were, O = Organisms per meter square
N = Average number of organisms per sample
A = Area of sampler in m^2
S = No. of samples taken at each station

(c) Fishes

Collection: Fishes were captured monthly from their natural habitat in the river Tawi with the help of cast net, drag net, gill net etc. The collected specimens were sorted at species level and all the species obtained were counted. The species abundance of sampling stations was determined by catch per unit effort (individual per net per hour). The data generated was used to examine spatial and temporal distribution patterns of different fish species at different study stations.

At every site, five nettings were attempted and captured fish specimens were transferred immediately into the river water filled tubs placed on the banks to reduce the chances of fish mortality and were counted. One or two specimens of each species were preserved in 10% formalin and were carried to laboratory for the purpose of identification (Day, 1878 and Jayaram, 1999). The remaining specimens were released back into the river at the same point from where they were captured. Catch per unit effort provides index of fish density and is directly proportional to the abundance of species (Biswas, 1993).

CPUE = No. of fish caught in one effort

(d) Macrophytes

Macrophytes inside the water were collected using an iron hook from predestined stations of the river by taking an area of (1x1) m^2 and placed them in a metallic sampler of known area whereas macrophytes slightly away from the banks or on the banks were collected by manual digging and cutting. Samples were taken to the laboratory for species identification.

RESULTS AND DISCUSSIONS

1. **Water quality:** A decrement in water quality parameters along the longitudinal profile of river Tawi was observed. Highest values of free carbon dioxide, calcium, magnesium, sulphates, phosphates, nitrates, chlorides, B.O.D were observed from downstream stations (St-III & IV), while lowest values of dissolved oxygen, transparency, water velocity were reported from St-III & IV. Lowest water velocity and higher depth was reported from downstream stations due to temporary impoundments. (Table-I) Decline in water quality parameters may be due to various processes associated with the construction such as dredging of sediments and their mixing in water column, channelization, accumulation of organic sewage and garbage in downstream stations due to low water velocity. Increase in turbidity and a reduction in dissolved oxygen content as a result of re-suspension of sediments caused due to dredging and physical disturbance to river bed was also reported by (Erftemeijer and Lewis 2006; Erftemeijer *et al.* 2012). A change in the colour, odour and a decrease in DO in disturbed sections of rivers have also been reported by Aazami *et al.* (2015).
2. **Sediment quality:** Sediment quality also showed a decline with highest sediment temperature, chloride, sulphate, phosphate and nitrate values were recorded from St-III & IV. pH showed a shift towards slightly acidic side. Downstream stations (St-III & IV) showed comparatively higher sediment calcium and magnesium content as compared to upstream stations which may be attributed to influx from agricultural run-off, sewage canals, drains, waste accumulation and decomposition at these stations. Present observation finds support in the findings of Verma and Saksena, (2010) and Sharma *et al.* (2010). A reduction in sediment quality parameters along the longitudinal profile of Tawi might be due to higher residence time of products of decomposition at these stations due to stagnation in isolated sections as a result of construction of barrage. In addition, mixing of sediments with the water column due to dredging and river bed eroding activities associated with channel deepening was responsible for release of some undesirable products products of decomposition from sediments into the overlying water column by way of re-suspension.
3. **Phytoplankton diversity and dynamics:** Phytoplankton population of river Tawi was represented by 12 genera viz. *Navicula sp. Pinnularia sps. Synedra sps. Cymbella sps. Gomphonema sps.* (Fam. Bacillariophyceae), *Cosmarium sp. Oedogonium sp. Scenedesmus sp. Spirogyra sp. Ulothrix sp. Volvox sp.* (Fam. Chlorophyceae) and *Oscillatoria sp.* (Fam.Cyanophyceae). While 12 genera belonging to all three

families were enlisted from upstream stations (St-I & II), downstream stations (St-III & IV) were found to be qualitatively poor with 7 and 9 species respectively during the first investigative year. During the 2ndyear, upstream stations maintained their comparatively higher diversity status with 11 species belonging to 3 families from St-I and 10 species belonging to 3 families at St-II, whereas only 06 species belonging to 2 families were reported from St-III and 8 species belonging to 2 families from St-IV. Family Cyanophyceae recorded its absence at St-III during both the years of study while it was absent at St-IV during 2ndyear of study. Family Bacillariophyceae was represented by a single genus *Navicula sps.* at the St-III. A considerable decline in quantitative aspect of phytoplankton was also seen along the longitudinal profile of river Tawi. Overall decrement in the density and diversity of phytoplankton from upstream (St-I & II) to downstream (St-III & IV) sections was observed which may be attributed to degraded water quality due to various point and diffuse, direct and indirect sources of pollution and the subsequent greater residence time of various products of decomposition at downstream sections due to pooled sections and reduction in riverine velocity, unfavourable habitat conditions due to repeated water diversions and irregular patterns of rise and fall in water levels due to construction of barrage (Somashekhar 1988; Wetzel 2001; Sawhney 2008 and Zohary and Ostrovsky, 2011). . While Bacillariophyceae showed strong positive correlations with transparency, pH, bicarbonate, calcium and magnesium, Chlorophyceae and Cyanophyceae established strong positive correlations with chloride, sulphate, phosphate and nitrate justifying the presence of blooms of Chlorophyceae in nutrient rich pooled sections of downstream stations.

4. **Zooplankton diversity and dynamics:** A total of 12 zooplankton genera belonging to four different groups were reported from river Tawi viz. *Centropyxis aculeata, Euglypha sps., Difflugia sps.* (Group-Protozoa), *Chydorus sps.* (Group-Cladocera), *Brachionus bidentatus, Brachionus calyciflorus, Brachionus quadridentatus, Keratella tropica, Euchlanisdilata, Philodina sps., Lecane bulla* (Group-Rotifera) and *Mesocyclops sps.* (Group-Copepoda). Group Rotifera dominated the zooplankton community of river Tawi both qualitatively (with a total of seven genera) and quantitatively at all four stations during both the years of study. Invariable presence of pollution indicator species such as *Brachionus sps.,Keratella sps.,*suggested a decline in water quality of river Tawi. Group Cladocera recorded complete absence from St-I, III & IV during both the years of study. During the 1st year, a total of 11 species were recorded at St-I, belonging to Protozoa (3 Species), Rotifera (7 species), Copepoda (1 genus) while during the 2nd year of study 10 species were recorded from this station, Protozoa (2 species), Rotifera (7 species) and Copepoda (1 genus). St-II was found to be qualitatively rich with representation from all four groups, with a total of 11 species were recorded from this station during the first year with Protozoa (2 species), Rotifera (7 species), Copepoda (1 genus), Cladocera (1 genus) and 12 species were recorded during the 2nd year with their distribution as Protozoa (3 Species), Rotifera (7 species), Copepoda (1 genus) and Cladocera (1 genus). At St-III, a total of 7 species belonging to Protozoa (1 genus), Rotifera (5 species), Copepoda (1 genus) were recorded at this station during the first year of study and 6 species Protozoa (1 genus), Rotifera (4 species), Copepoda (1 genus) were reported during the 2nd year of study. And lastly at St-IV, a total of 9 species

belonging to Protozoa (2 species), Rotifera (6 species), Copepoda (1 genus) were recorded during the first year of study while only 5 species Protozoa (1 genus), Rotifera (3 species), Copepoda (1 genus) were recorded during the 2nd year of study. A drastic decline in zooplankton diversity and quantitative abundance was observed along the longitudinal profile of river Tawi during the present investigation period (Aug., 2013-July, 2015) which may be due to activities associated with the barrage construction such as unpredictable or irregular habitat conditions resulting from repeated water diversions particularly at the Station-III, lead to sudden rise and fall in water levels and creation of dry land from previously inundated areas and vice versa, thus reducing the habitat suitability at the downstream stations, reduction in phytoplankton based food due to habitat degradation, high turbidity resulting from construction material, dredging, channelization, pollutant accumulation, plying of machinery and vehicles at the downstream stations and the physical alteration of the habitat conditions which collectively affect the water quality and thus causing a reduction in zooplankton diversity and abundance. While protozoa and copepoda showed strong positive correlations with air and water temperature, chloride, sulphate, phosphate, nitrate and B.O.D. Rotifers and cladocera showed the opposite trend justifying their peaks during winters. Activities associated with the construction of barrage such as dredging, deepening of river channel for the depth stipulated for the construction of reservoir or artificial lake by removal of deposited sediment at the downstream stations (St-III & IV) lead to the re-suspension of sediment in the water column resulting in increased turbidity which adversely affects the zooplanktonic population by decreasing light penetration and preventing oxygen diffusion (Seiyaboh *et al.,* 2013). Moreover, re-suspension of sediments or mixing of sediments in water column may result in release of toxic products of decomposition and heavy metals from the sediment into the water, thereby reducing the water quality and habitat suitability for the zooplanktons. Thorp and Mantovani, 2005, Rehman *et al,* 2016).

Unpredictable or irregular habitat conditions resulting from repeated water diversions particularly at the Station-III, lead to sudden rise and fall in water levels and creation of dry land from previously inundated areas and vice versa, thus reducing the habitat suitability at the downstream stations (Zohary and Ostrovsky, 2011). Accumulation of organic sewage, garbage, agricultural run-off and waste from the construction sites were found to render St-III & IV unfit for the survival of zooplankton (Wen *et al.,* 2011).

5. **Macrobenthic diversity and dynamics:** The macrobenthic fauna of River Tawi as enlisted in present study period are *Chironomous chironomous, Pentaneura sp., Tabanus sp., Culicoides sp., Eristalis sp., Psychoda sp., Micronecta sp., Berosus sp., Hydroglyphus sp.,* (Phylum-Arthropoda), *Tubifex tubifex., Pheretima posthuma, Erpobdella octoculata* (Phylum-Annelida) and *Physa acuta, Lymnaea acuminata and Gyraulus ladacensis* (Phylum-Mollusca). A total of 15 genera were enlisted with 9 genera belonging to Arthropoda and 3 genera from Annelida and 3 genera from Mollusca. An invariable and abundant presence of various pollution indicator species such as *Chironomous sp., Pentaneura sp., Physa sp, and Lymnaea sp.,* at all stations indicated an overall disturbed ecological status of river Tawi (Sharma, 2015). A general decline in macrobenthic invertebrate diversity was observed along

the longitudinal profile of river Tawi, with upstream St-II being the richest station qualitatively with 15 genera belonging to all three phyla during the 2nd year of study and St-III being the poorest with only 6 genera during both the years of study. While Station-I & II exhibited greater density and diversity of macrobenthic genera due to comparatively lesser anthropogenic disturbances to river bed, at the same time Station-III was found to be at the receiving end of various degrees of hydrological modifications such as frequent water diversions, dredging, channel deepening, in-constancy of flows leading to alternation of dry lands with inundated river bed during monsoon floods due to construction of barrage. Prolonged application of such disturbances and perturbations resulted in a decrease in the diversity of macrobenthos at this station. In addition to degraded water and sediment chemistry particularly lower dissolved oxygen and increased temperature at highly perturbed St-III affected macro-benthic fauna and their dynamics at this station. Furthermore, invariable presence of indicator species such as *Chironomous sps., Pentaneura sps., Tubifex sps., Physa sps.,* suggested a decrement in water and sediment quality at all four stations. Macrobenthic fauna of river Tawi depicted significant positive correlation with water chloride, sulphate, phosphate, nitrate, B.O.D content and a negative correlation with water velocity and dissolved oxygen content. Annelida and Mollusca showed strong positive correlation with water and sediment temperature while arthropoda showed negative correlation with temperature. In addition to degraded water and sediment chemistry, particularly lower dissolved oxygen and increased temperature in perturbed systems which affect macrobenthic fauna and their dynamics (Rader and Belish 1999; McKay and King 2006), distressed flow regimes are known to affect macro-invertebrates by altering physical habitat, interrupting their life history, limiting their longitudinal and lateral connectivity depending on the nature of the alteration, and facilitating the invasion and success of introduced species (Bunn and Arthington, 2002). Moreover, absence of macrophytes from the water column at St-III also resulted in decreased availability of diverse and morphologically structured microhabitats to macrobenthos (Borics *et al.,* 2003).

6. **Fish diversity and dynamics:** During the present investigative period of two consecutive years (August 2013 to July 2014 and August 2014 to July 2015) a total of 18 fish species belonging to five families i.e. Cyprinidae, Nemacheilidae, Mastacembelidae, Belonidae, and Bagridae were reported. Family Cyprinidae dominated the ichthyofauna of river Tawi and was represented by *Puntius conchonius, Puntius ticto, Puntius sophore, Labeo bata, Labeo boga, Tor tor, Tor putitora, Rasbora rasbora, Esomus danricus, Barilius bendelisis, Barilius vagra, Aspidoparia morar, Garra gotyla, Cyprinus carpio, Family Mastacembelidae by Mastacembelus armatus, Family Bagridae by Mystus bleekeri, Family Nemacheilidae by Nemacheilus botia, and Family Belonidae by Xenentodon cancila.* There was appreciable decrement in catch per unit effort and diversity along the longitudinal profile of River Tawi along with the normal seasonal changes. Station-I recorded the presence of a total of 16 fish species belonging to five families were reported from Station-I during the first year. All five families were represented at St-I with Cyprinidae (12 species), Family Mastacembelidae (1 species), Family Bagridae (1 species), Family Nemacheilidae (1 species), and Family Belonidae (1 species) and 14

species Cyprinidae (10 species), Mastacembelidae (1 species), Bagridae (1 species), Nemacheilidae (1 species) and Belonidae (1 species) were reported from the same station during the 2nd investigative year. A total of 17 fish species belonging to all five families were reported from Station-II during the first year with Cyprinidae (13 species), Family Mastacembelidae (1 species), Family Bagridae (1 species), Family Nemacheilidae (1 species) and Family Belonidae (1 species). While investigations during the 2nd consecutive year revealed the presence of 15 species belonging to five families; Cyprinidae (11 species), Mastacembelidae (1 species), Bagridae (1 species), Nemachelidae (1 species) and Belonidae (1 species) at this station. At Station-III, a total of 08 fish species belonging to two families Cyprinidae (7 species) and Nemacheilidae (1 species) were enlisted from this station during the 1st year of study whereas a total of 06 species belonging to a single family Cyprinidae were enlisted from this station during the 2nd year. At Station-IV, a total of 12 fish species belonging to three families viz Cyprinidae (10 species), Mastacembelidae (1 species) and Nemacheilidae (1 species) were reported during the 1st investigative year while a total of only 06 species belonging to a single family Cyprinidae were reported from this station during the 2nd year. Family Cyprinidae was represented at all four stations while Mastacembelidae, Belonidae, Bagridae, Nemacheilidae registered a total absence from downstream stations during 2nd year of study. Ichthyofauna of river Tawi depicted significant positive correlations with dissolved oxygen and transparency and negative correlations with air and water temperature and sulphate content. Since, increase in the velocity of river Tawi was always associated with significant increase in turbidity, thereby affecting the visibility, light penetration, phytoplankton mass and dissolved oxygen content, therefore water velocity depicted negative correlations with ichthyofauna of river Tawi. Nitrate and phosphate content depicted positive correlations with families Cyprinidae, Mastacembelidae, Bagridae and Belonidae depicting detritus and organic matter as important sources of food genera belonging to these families. Overall a decline in water quality parameters was associated with a decline in ichthyofaunal diversity at downstream stations (St-III & St-IV). Dissolved oxygen and temperature are known to affect the fish distribution on a larger scale (Moyle and Baltz, 1985) whereas current velocity and water depth influence fish distribution on a smaller scale in riverine systems (Bain *et al.,* 1988; Harvey and Stewart, 1991). Flow regime of a river is an important factor influencing fish distribution and assemblage structure (Stevens and Miller 1983, Meffe 1984, Poff and Allen 1995, Petersen and Kwak, 1999), and its modification by channelization, regulation, construction of dams and barrages etc. affects the fish assemblages at both the population and community levels (Schlosser 1991, Marchetti and Moyle 2001). Reduction in cover complexity due to constant disturbances to the river bed and homogenization of habitat was also found to be a major contributing factor to the decline in fish diversity and catch per unit effort in downstream stations. Role of cover at the scale of fish microhabitat was also advocated by Allouche (2002). Strong inter-relationship of fish production to the physical habitat variables of a river was also advocated by Bovee (1986) and Fausch *et al.* (1988). Feeding habits are also the strong determinants of survival in fluctuating conditions. Ample availability of macro-benthic invertebrate genera known to survive in low water and sediment quality conditions (Bloodworms, *Tubifex sps.*) which may form an important part of food of omnivorous species such as *Puntius sophore, Puntius ticto, Puntius conchonius* may have ensured their survival at all four stations. Important role of food availability and preferences in fish distribution and abundance was also advocated by (Filbert and Hawkins 1995, Hughes 1998, Nislow *et al.* 1999).

7. **Macrophyte diversity and dynamics:** During the present investigation period (Aug., 2013-July, 2015), River Tawi was found to be poor in macrophytic population at all four stations. Only five species of macro-phytes belonging to five different families and of two groups were reported from all the four stations viz. emergent macrophytes mostly confined to the banks or riparian areas *Colocasia esculenta* (Family-Araceae), *Scirpus sp.,* (Family-Cyperaceae) and *Ipomea carnea* (Family- Convolvulaceae and submerged-rooted *Potamogeton crispus* (Family-Potamogetonaceae) and *Hydrilla verticilla* (Family-Hydrocharitaceae). St-I & St-III were found to be completely devoid of any macrophytic population. A total of four species of macrophytes belonging to four different families viz. *Colocasia esculenta* (Family-Araceae), *Potamogeton crispus* (Family-Potamogetonaceae), *Scirpus sp.,* (Family-Cyperaceae) and *Hydrilla verticilla* (Family-Hydrocharitaceae) were reported from St-II. While *Colocasia esculenta* and *Scirpus sp.,* were reported in both consecutive years, other two species were seen in only second year of study at this station. *Ipomea carnea* recorded its constant presence at St-IV during both the years of study. Station-III was also devoid of any macrophytic population which may be attributed to construction of cemented embankments and shoreline developmental activities for the construction of barrage, water level fluctuations due to water diversions and extensive damage to riparian habitat. (Oglesby *et al.* 1976) While absence of any macrophytic species at St-I, might be due to the crematorium at the bank and the associated bank development activities. All the species observed from the river during the two-year study period were found to be hardy with an ability to survive in wide range of flow, pH, and temperature fluctuations along with having elaborate reproductive strategies (vegetative propagation), invasive and pollution tolerant which depicts the polluted and disturbed status of the river.

SUMMARY AND CONCLUSION

River Tawi was found to be at the receiving end of several habitat destructive activities such as dredging, channel modification, exposure and removal of sediments, disturbance to water column by various machines and vehicles deployed for construction, bank-development activities (hardening of river banks with concrete or construction of levee banks), dumping of construction waste (concrete etc.) in the river channel, water diversions, temporary impoundments and their subsequent conversion to fragmented marshy areas during the periods of low flows, accumulation of waste and associated reduction in water and sediment quality, altered sedimentation rate (increased sedimentation at downstream stations owing to stagnation and low velocity) etc. These disturbances resulted in altered channel width to depth ratios, decreased diversity and abundance of biota, decreased dissolved oxygen concentration with increasing water temperatures and overall reduction in water quality and loss of floodplain connectivity. Consequential responses of biota to different kinds of perturbations were apparent as loss in diversity, richness and abundance along with dominance of taxa tolerant to various kinds of disturbances (*Centropyxis sps.,*

Brachionus sps.,Chironomous sps., Pentaneura sps., Tubifex sps., Physa sps., Lymnaea sps.). Invariable presence of tolerant taxa at all stations depicted the stress on the riverine ecology by pollution but a sharp decrement in overall diversity and abundance at downstream sections as compared to upstream sections clearly depicted the localized accentuation of stress due to construction of barrage at downstream stations.

REFERENCES

- A.P.H.A. (1985). Standard methods for the examination of water and wastewater. (21st ed.). Washington: American Public Health Association.
- Aazami, J., Esmaili-Sari, A., Abdoli, A. (2015). Monitoring and assessment of water health quality in the Tajan River, Iran using physicochemical, fish and macroinvertebrates indices. *J Environ Health Sci Engineer* 13, 29.
- Adoni, A.D. (1985). Workbook on Limnology. Pratibha Publishers, C-10, Gour Nagar, Sagar, India.
- Allouche, S. (2002). Nature and functions of cover for riverine fish. *Bulletin Français de la Pêche et de la Pisciculture*, (365-366), 297-324.
- Bain, M.B., Finn, J.T. and Booke, H.E. (1988). Streamflow regulation and fish community structure. *Ecology*, 69(2), 382-392.
- Bender, E. A., Case, T. J. and Gilpin, M. E. (1984). Perturbation experiments in community ecology: theory and practice. *Ecology* 65, 1–13.
- Biswas, S.P. (1993). Manual of methods in fish biology. South Asian Publishers.
- Borics, G., Tothmeresz, B., Grigorszky, I., Padisak, J., Varbıro, G. and Szabo, S. (2003). Algal assemblage types of boglakes in Hungary and their relation to water chemistry, hydrological conditions and habitat diversity. *Hydrobiologia* 502: 145–155.
- Bovee, K. D. (1986). Development and evaluation of habitat suitability criteria for use in the instream flow incremental methodology. *U.S. Fish and Wildlife Service Biological Report* 86(7).
- Bunn, S.E. and Arthington AH. (2002). Basic principles and ecological consequences of altered flow regimes for aquatic biodiversity. *Environ Manage.* (4):492-507.
- Chopra, S.L. and Kanwar, J.S. (1967). Practical Agricultural Chemistry. S. Chand.
- Day, F. (1878). The fishes of India, Bernard Quaritch, London.
- Edmondson, W.T. and Winberg, G.G. (1971). A manual on productivity in freshwaters. *Blackwell Scientific Publications*, Oxford, p.358.
- Erftemeijer, P. L. A. Lewis, R. R. R. (2006). Environmental impacts of dredging on sea grasses: A review. *Marine Pollution Bulletin*, v. 52, p. 1553 – 1572.
- Erftemeijer, P. L. A., Riegl, B.; Hoeksema, B. W., and Todd, P. (2012). Environmental impacts of dredging and other sediment disturbances on corals: A review. *Marine Pollution Bulletin*, v. 64, p. 1737 – 1765.
- Fausch, K. D., C. L. Hawks, and M. G. Parsons. (1988). Models that predict standing crop of stream fish from habitat variables. USDA Forest Service General Technical Report PNW-GTR-213.

- Filbert, R. B., and C. P. Hawkins. (1995). Variation in condition of rainbow trout in relation to food, temperature, and individual length in the Green River, Utah. *Transactions of the American Fisheries Society* 124:824-835.
- Harvey B.C, and Stewart A.J. (1991). Fish size and habitat depth relationships in headwater streams: *Oecologia*, vol. 87(3), 336-342.
- Hughes, N. F (1998). A model of habitat selection by driftfeeding stream salmonids at different scales. *Ecology* 79: 281-294.
- Jayaram, K.C. (1999). The freshwater fishes of the Indian region.
- Lake, P.S. (2000). Disturbance, patchiness, and diversity in streams. *Journal of the North American Benthological Society*, 19:573–592.
- Marchetti, M. P. and Moyle, P.B (2001). Effects of flow regime on fish assemblages in a regulated California stream. *Ecological Applications* 11:530-539.
- McKay, S. and King, A.J. (2006). Potential effect of water extraction in small, unregulated streams. *River Research and Applications*, 22, 1023-1037.
- Meffe, G. K. (1984). Effects of abiotic disturbance on coexistence of predator-prey fish species, *Ecology* 65:1525-1534.
- Moyle P and Baltz, D. (1985). Microhabitat Use by an Assemblage of California Stream Fishes: Developing Criteria for Instream Flow Determinations. *Transactions of the American Fisheries Society*. 114. 695-704.
- Nislow, K. H., Folt, C.L and Parrish, D.L (1999). Favorable foraging locations for young Atlantic Salmon: application to habitat and population restoration. *Ecological Applications* 9:1085-1099.
- Oglesby RT, Vogel A, Peverly JH and Johnson R (1976). Changesin submerged plants at the south end of Cayuga Lake followingtropical storm Agnes. *Hydrobiol* 48: 251–255
- Opperman, J. J., S. Orr, H. Baleta, M. Dailey, D. Garrick, M. Goichot, A. McCoy, A. Morgan, L. Turley and A. Vermeulen. (2018). Valuing Rivers: How the diverse benefits of healthy rivers underpin economies. WWF.
- Pennak, R.W. (1978). Fresh water invertebrates of United States. 2nd edition. A Wiley interscience publication.
- Petersen, J. T. and Kwak, T.J. (1999). Modeling the effects of land use and climate change on riverine smallmouth bass. *Ecological Applications* 9:1391-1404.
- Poff, N. L., and Allen. J.D. (1995). Functional organization of stream fish assemblages in relation to hydrological variability. *Ecology* 76:606-627.
- Poff, N.L., Richter, B.D., Arthington, A.H., Bunn, S.E., Naiman, R.J., Kendy, E., Acreman, M., Apse, C., Bledsoe, B.P., Freeman, M.C., Henriksen, J., Jacobson, R.B., Kennen, J.G., Merritt, D.M., O'keeffe, J.H., Olden, J.D., Rogers, K., Tharme, R.E. And Warner, A. (2010). The ecological limits of hydrologic alteration (ELOHA): a new framework for developing regional environmental flow standards. *Freshwater Biology*, 55: 147-170
- Rader, R.B. and Belish, T.A. (1999). Influence of mild to severe flow alterations on invertebrates in three mountain streams. Regulated Rivers: *Research and Management* 15, 353-363.

- Rehman, M., Yousuf, A.R., Balkhi, M.H., Rather, M.I., Shahi, N., Meraj, M. and Hassan, K. (2016). Dredging induced changes in zooplankton community and water quality in Dal Lake, Kashmir, India. *African Journal of Environmental Science and Technology.* Vol. 10(5): 141-149.
- Sawhney, N. (2008). Biomonitoring of river Tawi in the vicinity of Jammu city. PhD Thesis, University of Jammu, Jammu.
- Schlosser, I. J. (1991). Stream fish ecology: a landscape perspective. *BioScience* 41:704-712.
- Seiyaboh, E.I., Inyang, I.R. and Gijo, A.H. (2013). Environmental impact of Tombia bridge construction across Nun River in central Niger Delta. *The International Journal of Engineering and Science* (IJES), Vol. 2(11): 32-41.
- Sharma, K.K. and Chowdhary, S. (2011). Macroinvertebrate assemblages as biological indicators of pollution in a Central Himalayan River, Tawi (J&K). *International Journal of Biodiversity and Conservation,* 3(5): 167-174.
- Sharma, K.K., Chowdhary, S. and Sharma, A. (2010). Malacofauna diversity of river Chenab fed stream (Gho-Manhasan). *The Bioscan,* 6(2): 267-269.
- Sharma, V. (2015). Sediment characterization of some sections of River Tawi and its impact on Macro-benthic faunal diversity. Ph.D Thesis, University of Jammu, Jammu.
- Somashekhar, R.K. (1988). Ecological studies of the two major rivers of Karnataka. In: Ecology and Pollution of Indian rivers. R.K Trivedi (ed.), Daya Publishing House, New Delhi.
- Sparks, R. E., Bayley, P. B., Kohler, S. L. and Osborne, L. L. (1990). Disturbance and recovery of large floodplain rivers. *Environmental Management* 14, 699–709.
- Stevens, D. E. and Miller, L.W (1983). Effects of river flow on abundance of young chinook salmon, American shad, longfin smelt, and Delta smelt in the Sacramento-San Joaquin River system. *North American Journal of Fisheries Management* 3:425-437.
- Thorp, J.H. and Mantovani, S. (2005). Zooplankton of turbid and hydrologically dynamic prairie rivers. *Freshwater Biology,* 50, 1474–1491.
- Tonapi, G.T. (1980). Freshwater invertebrates of India (An ecological approach).
- US EPA (2007). National Management Measures to Control Nonpoint Source Pollution from Hydromodification; Chapter 3: Channelization and Channel Modification. EPA 841-B-07-002 United States Environment Protection Agency.
- Verma, A.K. and Saksena, D.N. (2010). Influence of pollutants on bottom sediment of sewage collecting Kalpi (Morar) River, Gwalior, Madhya Pradesh (MP). *Journal of Toxicology and Environmental Health Sciences,* 2(4): 38-49.
- Ward, H.B. and Whipple, G.C. (1959). Freshwater Biology, 2nd edition. *John Wiley and Son.*
- Welch, P.S. (1952). Limnology: Mcg raw hill Book Company, New York, Toronto and London (2nd edition), 538.
- Wen, X., Xi, Y., Qian, F., Zhang, G. and Xiang, X. (2011). Comparative analysis of rotifer community structure in five subtropical shallow lakes in East China: role of physical and chemical conditions. *Hydrobiologia,* 661, 303–316.

- Wetzel, R.G. (2001). Limnology: Lakes and River Ecosystems, 3rd Ed. Published by: Academic Press, a Harcourt Science and Technology Company, New York, USA. Pp: 181-186.
- Williams, C.H. and Steinbergs, A. (1959). Soil sulphur fractions and chemical indices of available sulphur in some Australian soils. *Australian Journal of Agricultural Research,* 10: 340-352.
- Zohary, T. and Ostrovsky, I. (2011). Ecological impacts of excessive water level fluctuations in stratified freshwater lakes. *Inland Waters,* 1:1, 47-59.

CHAPTER

4

Faunal Diversity of the Great Indian Desert, India

Sanjeev Kumar[1] and Seema Kumar[2]
DRC, ZSI, Jodhpur[1] and AFRI, Jodhpur[2]
Email: dr.kumarsanjeev@gmail.com

ABSTRACT

The desert ecosystem is xeric in environmental physiology and dry in its character with extremely low rainfall and high evaporation. The desert landscape has dunal and interdunal feature with low flat lands as seasonally inundated wetlands also known as *Taal*. The present paper reports the faunal diversity of Great Indian Desert with especial reference to Rajasthan and Gujarat part of the Thar Desert. Total three thousand three hundred and forty six species have been recorded from the Great Indian Desert especially desert areas of Rajasthan and Gujarat. The present paper reports 2508 species of invertebrates and 769 species of vertebrates from the Great Indian Desert.

Key Words: Great Indian Desert, Fauna, Threatened species

INTRODUCTION

The deserts are unique diverse entities of the planet earth formed by mountains blocking the path of precipitation and thus share a common characteristic of low annual rainfall less than 25 cm. Deserts can be hot or cold but are always dry. Deserts falls in hot hyper-arid and arid regions with high temperatures in summers and sometimes below freezing temperatures in winters, high wind velocity, soils rich in minerals but with less organic matter and seasonal or dry riverbeds (Laity, 2008). Deserts are classified on the basis of temperature into four as hot temperate coasts, arid, hyper-arid and semiarid zones. India has both hot and cold deserts with a total area of about 127160 sq. km. The hot Thar Deserts spread over about 28500 sq. km between the Aravalli Hills in the north-east runs through Punjab and Haryana and the major portion lies in the north-west and the western part of Rajasthan, Gujarat, the Great Rann of Kachchh along the coast and the alluvian plains of the Indus River also. The Great Indian Desert has two important geographical features in the form of sand dunes and the Rann which are not only rich in biodiversity but also the most popular tourist's paradise from October to February world over. Located at the convergence of palaearctic and oriental regions, the origin of the Great Indian Desert is due to geo-tectonic and climate changes at the time of upheaval of the Himalaya and glaciations age and is an extension of Sahara desert all the way through Arabian and Persian deserts (Blanford, 1876, 1877; Gupta and Prakash, 1975). Most of the desert area is covered by huge shifting sand dunes, as the sand is highly mobile due to strong winds known as hot wind stroke (*loo*) before the onset of monsoon and receives sediments from the alluvial plains and the coast.

India is divided into ten bio-geographical zone including the Indian Desert Bio-geographic zone of 13 districts of Rajasthan and one district of Gujarat which covers an area of 215757 sq. km. (Rodger *et al.*, 2002) but the Great Indian Desert covers 13 districts of Rajasthan and 12 districts of Gujarat (Prakash 1996; Kumar, 2017) (Map 1, Map 2; Fig. 1, 2 and Table 1).

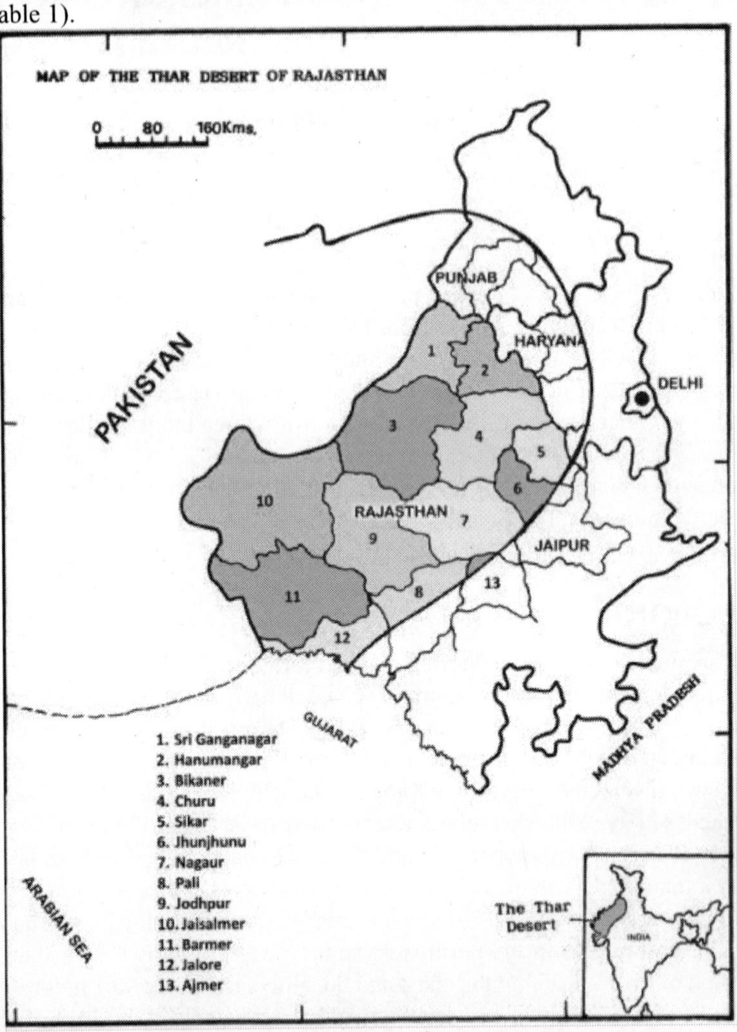

Map 1. *The Great Indian Desert, Rajasthan*

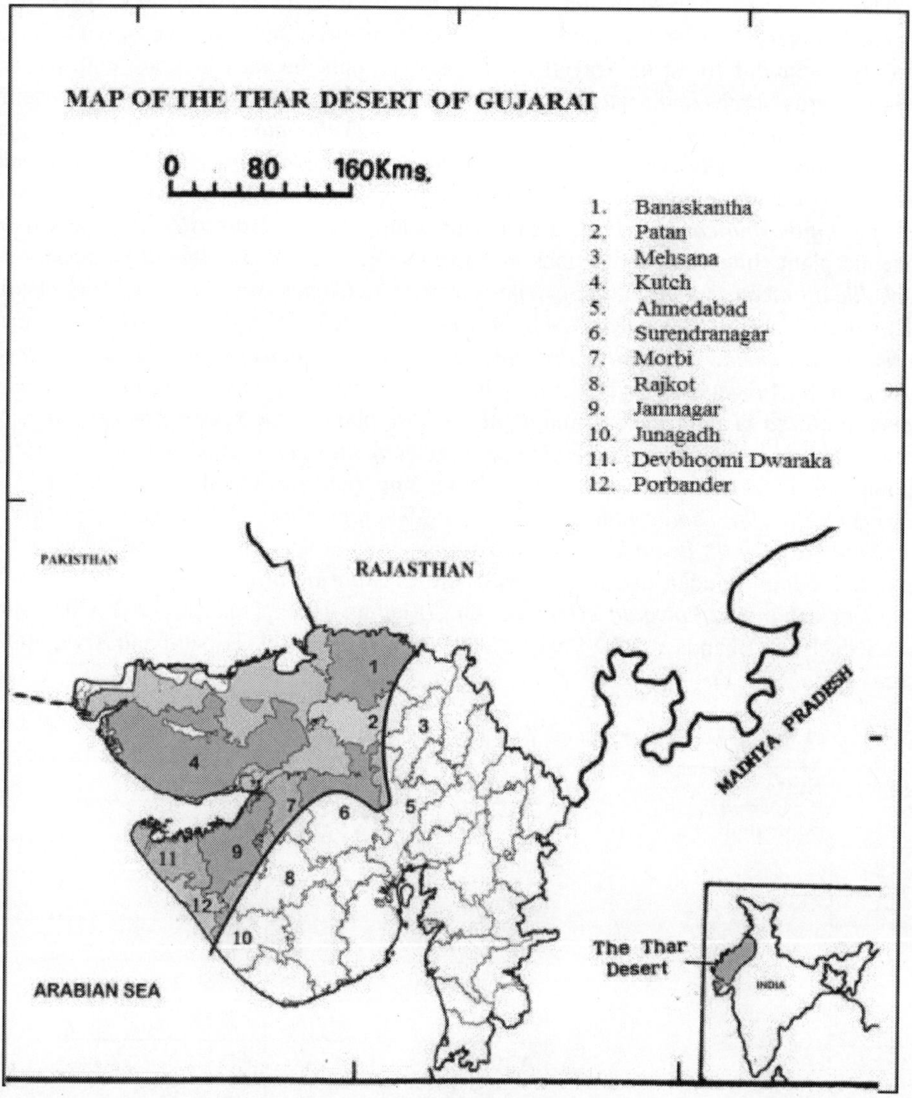

Map 2. *The Great Indian Desert, Gujarat*

The Great Indian Desert or the Thar Desert derives its name from the word thul meaning sand ridges or *thalis*. Large scale saline depressions are present in the form of Rann in Kachchh, Gujarat and some small saline patches in the Thar part of Taal-chhapar, Didwana, Pachpadra, Lunkaransar and Kuchaman in Rajasthan. The only important river is the Luni River which originates from *Naag Pahar*, near Puskar, Ajmer and finally drains into Rann of Kachchh. Though the climatic conditions are on the extreme side, the human population and the milch population is on the higher side with ever increasing trend and large scale development in these pristine ecosystems of desert areas. The *Gujjars*, *Rathis*, *Banjaras*, *Raika's*, *Rabaris* are important tribes of the region supporting the traditional values and practices of the dry environment.

The hot desert flora is mainly drought or salt tolerant (xerophytes), sparse but extremely diverse, comprises mainly of tropical thorn forest and scrub forest with blends into dry deciduous forest of Aravallis with seasonal ephemerals which are with thorns, stunted thorny shrubs and scattered trees. Some desert plants have long tap root systems due to which they survive. *Haloxylon salicornicum*, *Calligonum polygonoides*, *Tribulus rajastanensis*, *Crotalaria burhia*, *Salsola barvosma*, *Suaeda fruticosa*, *Aerva javanica*, *Apluda* sp. are important soil binders. *Catotropis procera* is a useful shrub for *katputli* making. *Euphorbia cordicifolia* is a succulent plant, whereas *Tephrosia falsiformis* is an endemic plant. Important indigenous and exotic tree species are *Acacia jacquemontii*, *Vachellia nilotica* (*Acacia nilotica*), *Senegalia senegal* (*Acacia senegal*), *Vachellia tortilis* (*Acacia tortilis*), *Balanites aegyptiaca, Salvadora persica, Salvadora oleoides, Capparis decidua, Tecomella undulata, Prosopis cineraria, Ziziphus nummularia, Moringa concanensis, Anogeissus sericea var nummularia, Tamarix aphylla, Lycium barbarum*. *Ephedra ciliata* is another very important and rare plant of the region. Invasive species such as *Prosopis juliflora, Parthenium sp. Argemone maxicana* are also spreading in the region and threat to native biodiversity. Other important medicinal plants are *Barleria prionitis, Citrullus colocynthis, Convolvulus microphyllus, Convolvulus prostratus, Commiphora wightii, Leptadenia pyrotechnicha*. *Atriplex* sp., are important annual herbs used as food and fodder. Important grasses are *Crotalaria burhia, Stipagrostis plumosa, Echinops echinatus, Polygala eriopteris*, etc. (Mathur, 1960; Prakash, 1963; Champion and Seth, 1968, Bhandari 1978; Shah, 1978; Goyal *et al*., 1999; Tripathi and Arya, 2002; Ishnava *et al*., 2011).

Table 1. *The zonation of the Thar Desert districts of Rajasthan and Gujarat*

Sl.No.	State	District	Areas in km^2
1.	Rajasthan	Pali	12387
2.		Jalor	10640
3.		Jodhpur	22850
4.		Sikar	7732
5.		Jhunjhunun	5928
6.		Nagaur	17718
7.		Ajmer	8481
8.		Sri-Ganganagar	10978
9.		Hanumangarh	9656
10.		Jaisalmer	38401
11.		Barmer	28387
12.		Bikaner	27244
13.		Churu	16830

14.	Gujarat	Kachchh	45652
15.		Jamnagar	8018
16.		Banaskantha	7621
17.		Devbhumi Dwaraka	5684
18.		Patan	4304
19.		Morbi	3409
20.		Surendranagar	2781
21.		Porbander	2294
22.		Ahmedabad	1434
23.		Mehsana	877
24.		Rajkot	755
25.		Junagadh	254

Fauna of the Great Indian Desert

The animals are able to adapt to live and survive in the deserts (xerocoles). The faunal diversity remains hidden during the day shine to control their body temperature and reduce their moisture requirements. Deserts are known by five S' *viz.* Sun, Sand, Salinity, Scorpions, and Snake. They are also known as land of reptiles characteristic being the Indian Spiny tailed lizards on which are dependant the avian raptors especially the winter migrants to the region. Though the harsh climatic conditions and sparse vegetation cover is there in the desert region but rich faunal diversity is represented by about 3346 species belonging to 17 major groups from protozoa to mammalian in and outside protected areas (PAs). The Kachchh Biosphere Reserve in Gujarat and Desert National Park are the two most important PAs. Sambhar Lake is an important wetland Ramsar site on the peripheral margin of the Great Indian Desert with wide spaces as corridors of faunal migration. The important wildlife sanctuaries (WLS) are Tal Chhapar WLS in Rajasthan and Indian Wild Ass Sanctuary, Kachchh Desert WLS, Kachchh Great Indian Bustard WLS, Narayan Sarovar Chinkara WLS in Gujarat. Besides there are 7 conservation Reserves in Rajasthan *viz.* Jor beed Gadwala, Bikaner; Gudha Bishnoiyan, Jodhpur; Gogelao, Nagaur; Rotu, Nagaur; Shakambari, Sikar & Jhunjhunu; Bir Jhunjhunu, Jhunjhunu and Khetri Bansyal, Jhunjhunu.

The fauna of Great Indian Desert has been studied by various researchers over the years by Stoliczka, 1872 a, b; Fraser, 1933, 1934, 1936; Ali, 1945 a, b, 1946, 1960, 1975; Abdulali, 1962, 1963; Daniel, 1963 a, b, 1975; Mathur and Mathur, 1963; Bhaskar, 1978; Johal & Dhillon, 1981; Mohnot, 1981; Thakur, 1981, 1989; Himmatsinhji, 1984; Mann, 1988; Gayan, 1999. In order to undertake detailed study on the faunal diversity of the desert areas of Rajasthan and Gujarat, scientists have worked out the fauna since 1960 which is documented in the form of different publications including contributions by scientists of ZSI, Jodhpur (Krishna & Prakash, 1960; Mathur & Mathur, 1963; Prakash, 1963, 1994; Ray & Mukherjee, 1969; Roonwal & Bose, 1969; Mahajan 1969, 1971, 1977 a, b; Mukherjee, 1970; Gupta & Gupta, 1976; Gupta & Prakash, 1975; Mahajan & Chandra, 1976; Mahajan & Mukherjee, 1977; Gupta , 1970, 1972, 1973, 1975, 1979; Gupta and

Gupta, 1976; Vazirani, 1973; Agrawal, 1967; Datta & Majumdar, 1970; Biswas, 1971; Deb, 1973; Gupta & Prakash, 1975; Bose & Mitra, 1976; Chhotani & Ray, 1976; Mahajan and Chandra, 1976; Mahajan and Mukherjee, 1977; Biswas and Sanyal, 1977; Ghosh, 1977; Sinha, 1980; Advani & Vazirani, 1981; Jakher *et al.* 1981; Sharma, 1981, 1982; Roonwal, 1983; Soota & Saxena, 1983; Prasad & Thakur, 1984; Mann, 1988; Joseph & Parui, 1992; Nandi, 1992; Varshney, 1992, 2002; Baqri & Jairajpuri, 1995; Mukherjee, 1995; Alfred & Agrawal, 1996; Ghosh *et al.*, 1996; Saxena, 1996, 2017; Tripathi *et al.* 1997; Vyas, 1998; Baqri, 2000; Chakraborty and Agrawal, 2000; Johal *et al.*, 2000; Chatterjee *et al.* 2004; Das, 2004; Kankane, 2000; Khanna, 2001; Alfred *et al.*, 2002; Banerjee and Mitra, 2002, 2004; Kazmi *et al.*, 2002; Kumar *et al.*, 2002, 2006; Purohit & Senacha, 2002, 2003, 2004; Kazmi & Ramamurthy, 2004; Rathore & Bhattacharrya, 2004; Saha *et al*, 2004; Baqri & Bohra, 2001, 2002 a, b, 2005; Senacha, 2003; Bohra & Baqri, 2004, 2008; Dookia & Tak, 2004; Sewak, 2004, 2009 a, b; Rathore & Tak, 2004; Tak & Rathore, 1996, 2004; Mandal and Hazra, 2005; Tyagi and Baqri, 2005; Kumar *et al.*, 2006; Pal *et al.*, 2006; Purohit *et al.*, 2006; Senacha, *et al.*, 2006; Hazra and Mandal, 2007; Kumar *et al.*, 2007; Julka *et al.*, 2009; Kumar & Gaur, 2009; Kumar & Kumar, 2009; Tak, 2009; Sewak 2009 a, b; Sharma 2009, 2013, 2015, 2018; Sivaperuman *et al.*, 2009; Sharma & Dhadeech, 2010; Sharma & Srivastava, 2010; Guevara & Sivaperuman, 2011; Bohra & Sultana, 2012; Mandal, 2012; Sharma & Kankane, 2012; Gaur *et al.*, 2013; Banyal & Kumar, 2013, 2014, 2016, 2017; Sharma *et al.*, 2013; Kankane *et al.*, 2014; Mukherjee *et al.*, 2014; Chhangani, 2015; Kumar & Kumar, 2015 a, b; Kankane & Kumar, 2016; Kumar, 2016, 2017; Sirawat, 2016; Kumawat & Purohit, 2020; Chandra *et al.*, 2020, 2021 and Kumar *et al.*, 2020). Important works on the protected areas have been in the form of publications (Editor-Director, 2000, 2004 a, b, 2005, 2009; Kankane *et al.*, 2014; Kumar & Banyal 2018).

Faunal Diversity of the Great Indian Desert: Gaps in Research

In an attempt to put together all the information about the known fauna of the Indian desert was made by the Zoological Survey of India and has reported about 2,043 species from single celled animal Protozoa to Mammalia from the entire Thar Desert of India which includes Rajasthan, Gujarat and parts of Punjab and Haryana (Gosh *et al.,* 1996, Sharma, 2013) till early nineteen's which was further updated based on the research carried out and review of literature. In all about 3346 species of different faunal taxa belonging to 1901 genera and 633 families under 17 phylum have been reported (Kumar, 2017; Chandra *et al.*, 2021). From the Great Indian Desert about 3156 species are reported from Rajasthan and 894 species are from the Gujarat part. The invertebrate groups are the dominant group represented by 2508 species, followed by vertebrates represented by 769 species and least reported are the protozoans represented by 69 species only (Fig 1.).

The desert areas bordering Aravallis has transitional vegetation with rarefied environment and affected by marginal impacts of climatic undulation as also faunal species migration. The district of Sri Ganganagar, Hanumangarh, Bikaner, Jaisalmer, Barmer, Jodhpur and Churu are entirely desertic in nature whereas districts such as Nagaur, Ajmer, Pali, Jalore, Jhunjhunu and Sikar are partially fall under the ambit of desert ecosystem Map 1. Infact, district like Sirohi is also just embraced by the marginal belt of the Thar Desert and therefore reflects the desertic character. The district of Ajmer also has a very small portion falling under the desert ecosystem but it has passages through Aravallis traversing the ecosystem properties across the two realms of arid and semi-arid regions.

Large open galleries of spaces across Aravallis allows frequent migration of faunal species and admixture of such biota can be observed vividly in this region.

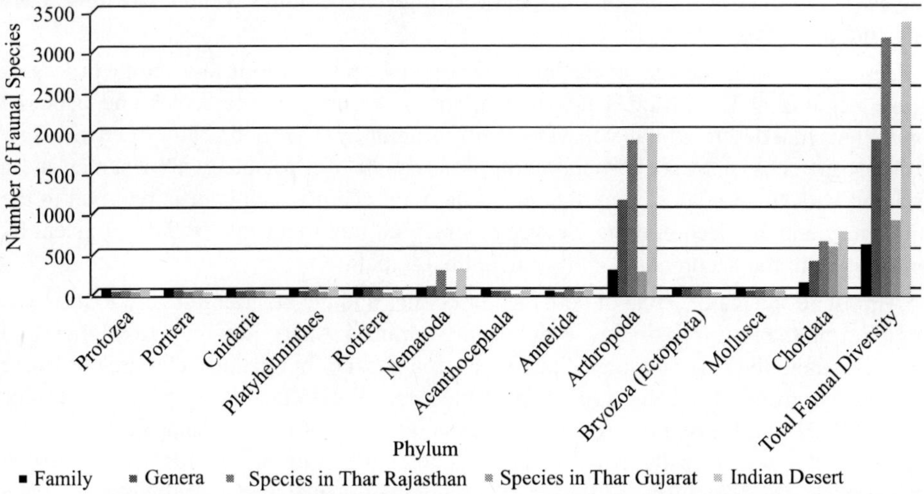

Fig 1. *Taxonomic status of Faunal Diversity in the Great Indian Desert*

Similarly, southern boundary of the Thar Desert lies along Great Rann of Kachchh and sandy plain and fossil dune systems of northern Gujarat. Total twelve districts of Gujarat comes under desert region *viz.* Banaskantha, Patan, Kachchh, Mehsana, Ahmedabad, Surendranagar, Morbi, Rajkot, Jamnagar, Junagadh, Porbander and Devbhumi Dwaraka (Table 1.). Of these Kachchh, Porbander and Devbhumi Dwaraka are fully desert districts and only some parts of the remaining districts are regarded as desert.

Deserts support diverse communities of plant and animals that have evolved over the period and are well adapted for their survival by various protective devices against the extreme temperatures and arid conditions. Invertebrates, particularly arthropods such as flies, beetles, ants, termites, locusts, millipedes, scorpions and spiders have hard waxy cuticles which prevent water from escaping and many of them lay their eggs underground and their young develop away from the temperature extremes at the surface. Amphibians of desert regions are mainly fossorial, spending the hot dry months aestivating in deep burrows. Reptiles can withstand the extreme temperatures by control of their body temperatures, some change colours and have sharp scales for defence, and others change their appearance to look more threatening. The avian species like sandgrouse has special feathers that soak up water which helps them to carry the water to its young in the nest. Many desert mammals are burrowers.

Many other animals in hot desert are nocturnal, seeking out shade during the day or dwelling underground in burrows or undergo aestivation. Diurnal faunal species protect themselves from the heat by spending most of the day in whatever shade available in the desert scenario. Thick fur coat and small mammals as well as several lizards, all dig burrows to wait out the cold weather. Camouflage is another important feature of desert fauna like sandgrouse and the desert lark takes frequent dust baths which ensures that it matches its environment. Herbivorous obtain moisture from the plants they eat. Carnivores fulfill their water requirement from the blood or body fluids of their prey. The desert fauna is incomplete if mention is not made of dromedary camels, a lifeline of desert people as they use very little water and can withstand very high temperatures without sweating. They also store fat in their humps for food.

Desert ecology is the sum of the interactions between both biotic and abiotic processes in arid regions and it includes the interactions of xerophytes, xerocoles and bacterial populations in a desert habitat, ecosystem and community. The food chains or energy flow in the desert ecosystem is short and comprises mainly of a producer, a consumer and a predator, with the predator being the top of the food chain which later is returned to the soil as nutrients by decomposers. Scavengers play an important role in the ecosystem by contributing to the decomposition of dead animal remains.

Almost all the major phyla of both the vertebrates and the invertebrates are found here, ranging from the tiny, microscopic Protozoa to Mammals. Apart from major exploration of reptilian faunal diversity, among the mammal's rodents, chiroptera, primates are one of the well-studied group in the Rajasthan part of the Great Indian Desert (Purohit & Senacha, 2002, 2003, 2004; Purohit *et al.*, 2006; Senacha, *et al.*, 2006). The Indian desert fauna is extremely rich in species diversity of mammals and winter migratory birds. Even breeding ground of Greater Flamingo and Lesser Flamingo other than at the Great Rann of Kachchh in India was discovered by the scientists of ZSI from Sambhar Lake, Rajasthan.

The impact of climatic change is very well recorded in the case of outbreak of desert locust *Schistocerca migratoria* (Kumar *et al.*, 2020) and the decline in population and habitat shrinkage in the case of Great Indian Bustard *Ardeotis nigriceps* (Vigors, 1831) (Kumar 2020).

Some important vertebrate species of the desert are Desert Monitor *Varanus griseus* (Daudin, 1803), Spiny tailed lizard - *Sara hardwickii* Gray, 1827, Central Asian Cobra - *Naja oxiana* (Eichwald, 1831); Great Indian Bustard *Ardeotis nigriceps* (Vigors, 1831), Houbara Bustard *Chlamydotis undulata* (Jacquin, 1784), Demoiselle Crane *Grus virgo* (Linnaeus, 1758), Black-bellied Sandgrouse *Pterocles orientalis*, White-browed Bushchat *Saxicola macrorhynchus* (Stoliczka, 1872), Cream-coloured Courser *Cursorius* cursor Latham, 1787; Caracal *Caracal caracal* (Schreber, 1776), Desert Cat *Felis silvestris* Schreber, 1777, Desert Fox *Vulpes vulpes pusilla* Blyth, 1854, Chinkara *Gazella bennettii* (Sykes, 1831), Scaly Anteater *Manis crassicaudata* Gray, 1827, Wolf *Canis lupus pallipes* Sykes, 1831, etc.

Gaps in research still exists in the knowledge on the faunal diversity known from the Great Indian Desert especially the invertebrate groups. So far only 3.3 % of the Indian faunal diversity is known from the Indian Desert Zone (Fig. 2, Fig. 3, Fig. 4). The phylum Protozoa, Porifera, Cnidaria, Platyhelminthes, Rotifera, Acanthocephala, Annelida, Bryozoa are represented only from Rajasthan part of the Great Indian Desert and no species are reported from the Gujarat part. Groups like Mesozoa, Ctenophora, Gastrotricha, Kinorhyncha, Spiuncula, Echiura, Onychophora, Phoronida, Entoprocta, Bracghiopoda, Chaetognatha, Tardigrada, Nemertea, Echinodermata, Hemichordata are not represented from the Great Indian Desert due to the desertic conditions. The mollusca are represented by only 36 species including 24 freshwater and 12 terrestrial species.

The Chordata are represented by different groups accounting to about 769 species represented by Fishes, Amphibia, Reptiles, Aves and mammals (Fig. 5). As per IUCN (2020) about 54 species of vertebrates are under various categories of threatened species from the Great Indian Desert as compared to 579 threatened species recorded from India (Table 2 & Table 3). The important Endangered species from Great Indian Desert are 3 species of fishes *viz. Tor putitora* (Hamilton, 1822) (Rajasthan); *Labeo potail* (Sykes, 1839) (Banaskantha, Surendranagar Gujarat) and *Dawkinsia arulius* Jerdon, (1849) (Surendranagar, Gujarat); one species of Amphibia *Minervarya sahyadris* (Annandale, 1919) from Kachchh, Gujarat; two species of reptiles are Critically endangered *viz. Batagur baska* (Gray, 1831) from Kachchh, Gujarat and *Batagur kachuga* (Gray, 1831) from Rajasthan whereas 8 species of aves are Critically Endangered the most important being the Great Indian Bustard *Ardeotis nigriceps* (Vigors, 1831) which is near extinction and only restricted to Jaisalmer district, Rajasthan and Kachchh district Gujarat (Kumar, 2019). One species of mammals *viz. Manis crassicaudata* Geoffroy, 1803 is Endangered and two species viz. *Lutrogale perspicillata* (Hilaire, 1826) and *Panthera pardus* (Linnaeus, 1758) are under Vulnerable category (IUCN, 2020). The list of threatened vertebrate fauna from Rajasthan and Gujarat part as compared to Indian scenario is given in Table 2.

Table 2. *Threatened Species of Vertebrate Fauna from India w.e.r. to the Great Indian Desert.*

Class	Critically Endangered (CR)		Endangered (EN)		Vulnerable (VU) / Near Threatened (NT)		Overall Threatened	
	India	Thar Desert	India	Thar Desert	India	Thar Desert	India	Thar Desert
Fishes (Pisces)	32	0	90	3	125	5	247	8
Amphibia	20	0	36	1	26	0	82	1
Reptilia	10	2	18	0	26	3	54	5
Aves	17	8	21	10	63	19	101	37
Mammals	9	0	40	1	46	2	95	3

Table 3. *List of Threatened Species from the Great Indian Desert, Rajasthan and Gujarat*

Class	Order	Family	Species	Threatened Status (IUCN)	Distribution	
					Rajasthan	Gujarat
Fishes (Pisces)	Cypriniformes	Cyprinidae	*Cirrhinus cirrhosus* (Bloch, 1795)	VU		+
			Cyprinus carpio Linnaeus, 1758	VU	+	+
			Labeo potail (Sykes, 1839)	EN	+	+
			Dawkinsia arulius Jerdon, (1849)	EN	+	+
			Tor putitora (Hamilton, 1822)	EN	+	

	Channiformes	Channidae	*Channa orientalis* Bloch & Schneider, 1801	VU	+	+
	Siluriformes	Siluridae	*Wallago attu* (Bloch and Schneider, 1801)	VU	+	+
	Perciformes	Cichlidae	*Oreochromis mossambicus* (Peters, 1852)	VU		+
Amphibia	Anura	Dicroglossidae	*Minervarya sahyadris* (Annandale, 1919)	EN		+
Reptilia	Testudines	Geoemydidae	*Batagur baska* (Gray, 1830)	CR		+
			Batagur kachuga (Gray, 1831)	CR	+	
		Testudinidae	*Geochelone elegans* (Schoepff, 1795)	VU	+	+
	Crocodylia	Crocodylidae	*Crocodylus palustris* Lesson, 1831	VU		+
	Squamata	Pythonidae	*Python molurus* (Linnaeus, 1758)	NT	+	+
Aves	Otidiformes	Otididae	*Ardeotis nigriceps* (Vigors, 1831)	CR	+	+
			Chlamydotis undulata (Jacquin, 1784)	VU	+	+
			Chlamydotis macqueenii (Gray, 1832)	VU	+	+
			Sypheotides indicus (Miller, 1784)	EN	+	+
	Anseriformes	Anatidae	*Anser erythropus* (Linnaeus, 1758)	VU		+
			Aythya ferina (Linnaeus, 1758)	VU	+	+
			Aythya baeri (Radde, 1863)	CR	+	+
			Marmaronetta angustirostris (Ménétriés, 1832)	VU	+	+
	Gruiformes	Gruidae	*Antigone antigone* (Linnaeus, 1758)	VU	+	+
			Leucogeranus leucogeranus (Pallas, 1773)	CR		+
	Ciconiiformes	Ciconiidae	*Ciconia episcopus episcopus* (Boddaert, 1783)	NT	+	+
			Leptoptilos javanicus (Horsfield, 1821)	VU	+	+
			Leptoptilos dubius (Gmelin, 1789)	EN	+	+

	Charadriiformes	Laridae	*Rynchops albicollis* Swainson, 1838	EN	+	+
			Sterna acuticauda Gray, 1832	EN		+
		Charadriidae	*Vanellus gregarius* (Pallas, 1771)	CR	+	+
		Scolopacidae	*Calidris tenuirostris* (Horsfield, 1821)	EN		+
			Gallinago nemoricola Hodgson, 1836	VU		+
			Tringa guttifer (Nordmann, 1835)	EN	+	+
	Columbiformes	Columbidae	*Columba eversmanni* Bonaparte, 1856	VU	+	+
	Accipitriformes	Accipitridae	*Gyps indicus* (Scopoli, 1786)	CR	+	+
			Gyps tenuirostris Gray, 1844	CR	+	+
			Gyps bengalensis (Gmelin, 1788)	CR	+	+
			Neophron percnopterus (Linnaeus, 1758)	EN	+	+
			Sarcogyps calvus (Scopoli, 1786)	CR	+	+
			Aquila nipalensis Hodgson, 1833	EN	+	+
			Aquila heliaca Savigny, 1809	VU	+	+
			Aquila rapax (Temminck, 1828)	VU	+	+
			Clanga clanga (Pallas, 1811)	VU	+	+
			Clanga hastata (Lesson, 1831)	VU	+	+
			Haliaeetus leucoryphus (Pallas, 1771)	EN	+	+
	Falconiformes	Falconidae	*Falco cherrug* Gray, 1834	EN		+
	Passeriformes	Estrildidae	*Amandava formosa* (Latham, 1790)	VU	+	+
		Paridae	*Machlolophus nuchalis* (Jerdon, 1845)	VU	+	+
		Muscicapidae	*Saxicola macrorhynchus* (Stoliczka, 1872)	VU	+	+
		Locustellidae	*Chaetornis striata* (Jerdon, 1841)	VU		+
		Laniidae	*Lanius meridionalis* Temminck, 1820	VU	+	+

Mammalia	Pholidota	Manidae	*Manis crassicaudata* Geoffroy, 1803	EN	+	+
	Carnivora	Mustelidae	*Lutrogale perspicillata* (Geoffroy Saint-Hilaire, 1826)	VU	+	+
		Felidae	*Panthera pardus* (Linnaaeus, 1758	VU	+	+

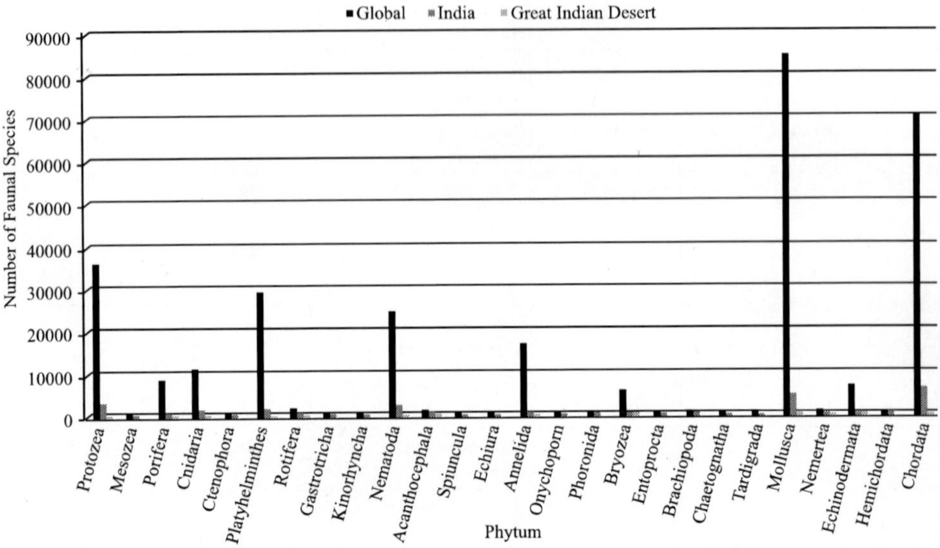

Fig 2. *Taxonomic Status of Faunal Diversity of Great Indian Desert w.e.r. to India and the world*

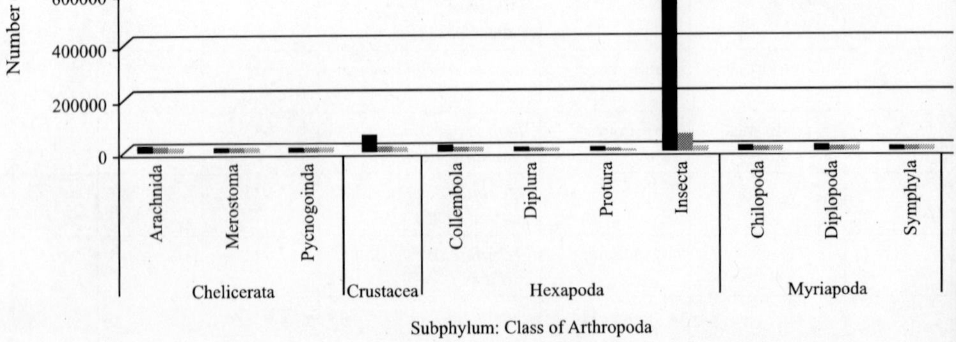

Fig. 3. *Taxonomic Status of Invertebrate Faunal Diversity: Arthropoda of Great Indian Desert*

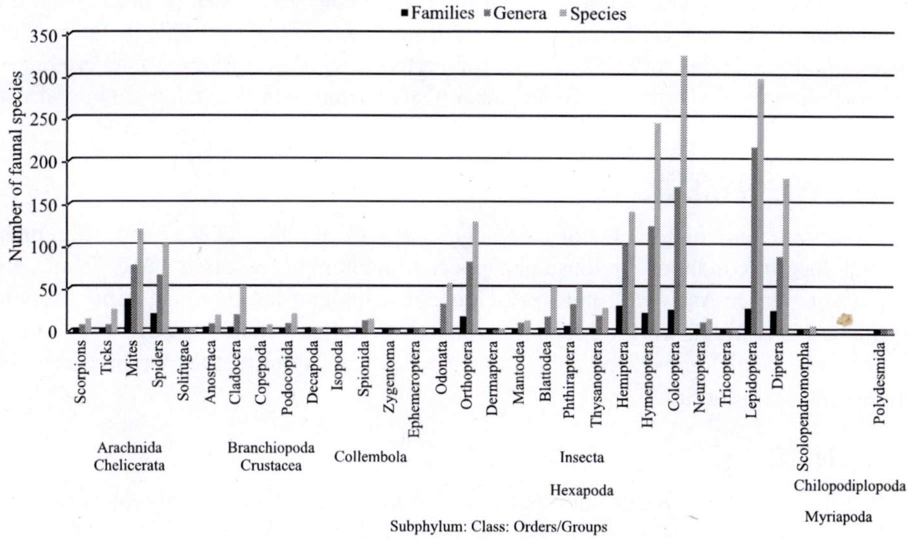

Fig. 4. *Detail Taxonomic status: Arthropoda of Great Indian Desert*

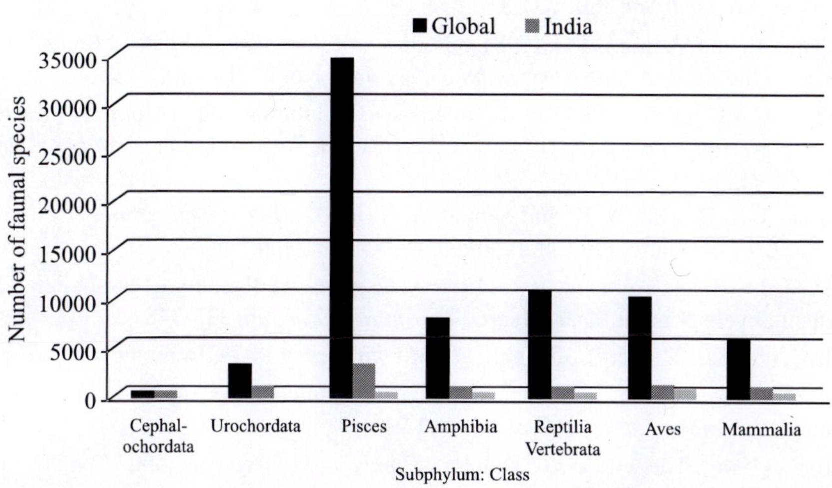

Fig. 5. *Taxonomic status of Chordata w.e.r. to Vertebrate Faunal Diversity of Great Indian Desert*

CONCLUSION

Fauna of desert districts of Rajasthan (Sri Ganganagar, Hanumangarh, Bikaner, Jaisalmer, Brmer, Jodhpur, Churu, Nagaur, Ajmer, Pali, Jalore, Jhunjhunu and Sikar) and Gujarat (Banaskantha, Patan, Kutch, Mehsana, Ahmedabad, Surendranagar, Morbi, Rajkot, Jamnagar, Junagadh, Porbander and Devbhumi Dwaraka) is compiled in this paper, based on all the existing literature data as far as possible and survey work done over the last 28 years.

Total three thousand three hundred and forty-six species have been recorded from the Great Indian Desert especially desert areas of Rajasthan and Gujarat. This includes 2508 species of invertebrates and 769 species of vertebrates. Many groups of invertebrates still remain unexplored from the Great Indian Desert from both the states of Gujarat and Rajasthan.

ACKNOWLEDEMENTS

The authors are thankful to Directors, Zoological Survey of India, Kolkata and Arid Forest Research Institute, Jodhpur for constant support throughout the research work. Thanks are due to Professor Dr. Ashok Purohit, JNV University, Jodhpur for giving an opportunity to contribute this research work on faunal diversity of the Great Indian Desert. The authors are also grateful to all those who helped us immensely in exploring the literature from the libraries and various other sources.

REFERENCES

- Abdulali, H. (1962). An Ornithological trip to the Gulf of Kutch. *J. Bombay nat, Hist. Soc.,* 59: 655-8.
- Abdulali, H. (1963). Ornithological notes on a second trip to the Gulf of Kutch. *J. Bombay nat, Hist. Soc*., 60: 703-8.
- Advani, R. and Vazirani, T.G. (1981). Studies on ectoparasites of bats of Rajasthan and Gujarat (India), *Rec. zool. Surv. India, Occ. paper,* No. 22, 155 pp.
- Agrawal, V.C. (1967). Field observations on the biology and ecology of the Desert Gerbil, *Meriones hurrianae* (Rodentia: Muridae) in Western India. *J. zool. Soc. India,* 17 : 125- 134.
- Alfred, J. R. B., Das, A. K. and Sanyal. A. K. (2002). *Ecosystems of India*. Published by the ENVIS Centre, Zool. Surv. India, Kolkata. 410 pp.
- Alfred, J.R.B. and Agrawal, V.C. (1996). The mammal diversity of the Indian desert. Faunal Diversity in the Thar Desert: *Gaps in Research,* pp. 335-348.
- Ali, S. (1945a). The birds of Kutch. Oxford University Press, Bombay.
- Ali, S. (1945b). More about the flamingo [*Phoenicopterus ruber rosus* Pallas]. in Kutch. *J. Bombay nat, Hist. Soc.,* 45: 580-93.
- Ali, S., (1946). The wild ass of Kutch. *J. Bombay nat. Hist. Soc.*, 48: 472-477.
- Ali, S. (1960). The Cinereous Vulture [*Aegypius monachus* (Linnaeus)] an addition to the birds of Kutch. *J. Bombay nat. Hist. Soc.,* 57: 408-409.
- Ali, S (1975)..On some brids of the Indian Desert. In *Environmental Analysis of the Thar Desert.* (Ed. R. Gupta and I. Prakash), pp. 423-431, 1 pl. English Book Depot, Dehra Dun.
- Banerjee, D. and Mitra, B. (2002). Beeflies (Bombyliidae: Diptera) of the Thar Desert. *Insect Environmental.* 7 (4): 190-191.
- Banerjee, D. and Mitra, B. (2004). Some Bee-flies (Bombyliidae: Diptera) from Thar Desert. *Records of the Zoological Survey of India*, 103(1-2): 67.
- Banyal, H.S. and Kumar, S. (2013). Fish diversity of seasonal wetlands of Kachchh

Biosphere Reserve (KBR), beyond India bridge, Gujarat. *J. Env. Bio-Sci.*, 27(1): 129-130

- Banyal, H.S. and Kumar, S. (2014). Ichthyofaunal Diversity of Tundi Wetland, Bajana, Little Rann of Kachchh, Gujarat, India. *J. Bombay nat. Hist. Soc.*, 111(2): 140-144.
- Banyal, H.S. and Kumar, Sanjeev (2016). Additions to the Ichthyo-faunal Diversity in Luni River near Luni Town, Jodhpur (Rajasthan). *Bionotes,* 18 (4): 123-124.
- Baqri, Q.H. (2000). Diversity in plant and soil nematodes (Nematoda) of Rajasthan (India): Gaps in Research. In *Plant Diseases* (edited by P.C. Trivedi); Pointer Publishers, Jaipur: 286-304.
- Baqri, Q.H. and Bohra, P. (2001). Nematodes from Rajasthan, India. I. Six new species of Dorylaimida. *Nematology,* 3 (2): 113-127.
- Baqri, Q.H. and Bohra, P. (2002a). Deserts: Rann of Kachchh. Pp. 111–122. In: Alfred, J.R.B., A.K. Das and A.K. Sanyal (Eds): Ecosystems of India. *ENVIS*, Zoological Survey of India, Kolkata.
- Baqri, Q.H. and Bohra, P. (2002b). Diversity of Plant and Soil Nematodes in the Fragile Ecosystem of Indian Thar Desert. In: Nematode Diversity (Ed. M.S. Jairajpuri). Silverline, P.S. Nagar, Hyderabad: 336-358.
- Baqri, Q.H. and Bohra, P. (2005). Studies on Plant and Soil Nematodes associated with crops of economic importance in Gujarat. *Rec. zool. Surv. India, Occ. Paper* No., 240: 1-48.
- Baqri, Q.H. and Jairajpuri, M.S. (1995). Bibliography of Nematology of India. CBS Publishers & Distributors, Daryaganj, New Delhi, 279 pp.
- Bhandari, M.M. (1978). The Flora of Indian Desert Scientific Publisher, Jodphur, India.
- Bhaskar, S. (1978). Notes from the Gulf of Kutch. *Hamadryad,* 4(l): 3-6.
- Biswas, S. (1971). Fauna of Rajasthan, India. Part 11. Crustacea: Cladocera. *Records of Zoological Survey of India,* Delhi, 63 (1965]: 95-141.
- Biswas. S. and Sanyal. D.P. (1977). Fauna of Rajasthan. India. Part: Reptilia. Records of Zoological Survey of India, Delhi. 73 [1975]: 247-269.
- Blanford, W.T. (1876). On the physical geography of the Great Indian Desert with special reference to the former existence of the sea in the Indus Valley: and on the origin and mode of formation of sand hills. *Journal of Asiatic Society of Bengal,* Calcutta, 45: 86-103.
- Blanford, W.T. (1877). Geological notes on the Great Indian Desert between Sind and Rajputana. *Records of Geological Survey of India,* Calcutta, 10: 10-21.
- Bohra, Padma and Sultana, Razia. (2012). Nematodes of Thar Desert, Rajasthan. *Rec. zool. Surv. India, Occ. Paper No.,* 334: 1-95, (Published by the Director, *Zaol. Surv. India,* Kolkata).
- Bohra, P. and Baqri, Q.H. (2004). Nematoda, In: State Fauna Series No.8, Gujarat (Part 2); Edited by the Director, Zoological Survey of India, Kolkata. pp. 355-400.
- Bohra, P. and Baqri, Q.H. (2008). Addition to the fauna of plant and soil nematodes of Gujarat, india. *Rec. zool. Surv. India.* l08 (Part-4): 5-15.
- Bose, G. and Mitra, T.R. (1976). The Odonata fauna of Rajasthan. *Records of Zoological Survey of India,* Delhi, 71 [1973]: 1-11.

- Chhangani, A.K. (2015). Population ecology of Vultures in the western Rajasthan, India. *Indian forester*. 131 (10):1373-1382.
- Chakraborty, S. and Agrawal, V.C. (2000). Mammalia. In: State fauna series No. 8 – Fauna of Gujarat (Part 1), pp. 15-83. Published Director, ZSI, Kolkata.
- Champion, H.G. and Seth, S.K. (1968). A revised survey of the forest types of India. Manager Publ., Delhi. 1-73.
- Chandra, K., Raghunathan, C. and Sheela, S. (2020). *Animal Discoveries 2019: New Species New Records*: 1-180 (published by the Director, Zool. Surv. India, Kolkata).
- Chandra, K., Gupta, D., Raghunathan, C., Kumar, S. and Saini, J. (2021). Faunal Diversity of Biogeographic Zones of India: Desert: 1-158. (Published by the Director, ZSI, Kolkata).
- Chatterjee, S.K., Chakraborty, S., Biswas, D.N., Halder, S.K., Ghosh, S.K., Mukhopadhyay, P. and Biswas, S. (2004). Insecta: Coleoptera. In: State Fauna Series No. 8. *Fauna of Gujarat*: Invertebrates (Part 2): 95-103. Published by Director, ZSI, Kolkata
- Chhotani, O.B. and Ray, K.K. (1976). Fauna of Rajasthan, India. Hymenoptera. *Records of Zoological Survey of India,* Delhi. 71 [1973]: 13-49.
- Daniel, J.C. (1963a). Field guide to amphibians of western India. Part I. *J. Bombay nat. His. Soc.,* 60: 415-438.
- Daniel, J.C. (1963b). Field guide to amphibian of western India. Part II. *J. Bombay nat. His. Soc.,* 60: 690-702.
- Daniel, J.C. (1975). Field guide to amphibian of western India. Part III. *J. Bombay nat. His. Soc.,* 72: 506-522.
- Das A. K. (2004). Protozoa. *State Fauna Series* No.8 – Fauna of Gujarat (Part 2), pp. 417-427. Eds. Director, Zoological Survey of India, Kolkata.
- Gosh, A.K. Baqri, Q.H. and Prakash, I. (1996) (Eds.). *Fauna in the Thar Desert. Faunal diversity of Thar. Gaps in Researc*h. Scientific Publishers.
- Datta, A.K. and Majumdar, N. (1970). Fauna of Rajasthan. India. Part 7. Fishes. Records of Zoological Survey of India, Delhi, 62 [1964]: 63-100, 9 pls.
- Deb, M. (1973). Fauna of Rajasthan (India): Crustacea-Ostracoda. *Rec. zool. Surv. India*, 67[1969]: 233-259.
- Dookia Sumit and Tak Jeeva Ram (2004). Status and distribution of Indian Flying Fox (*Pteropus gigantius* Brunnich) in Thar Desert of Rajasthan. BAT NET - CC1NSA Newsletter, 5: 23-24.
- Editor-Director. (2000). Fauna of Gujarat (Part 1). *State Fauna Series* No. 8 Published, Director, Zoological Survey of India, Kolkata.
- Editor-Director. (2004a). Fauna of Gujarat (Part 2). *State Fauna Series* No. 8 – (Published, Director, *Zool Surv India*, Kolkata).
- Editor-Director. (2004b). Fauna of Desert National Park, Rajasthan (Proposed Biosphere Reserve). *Conservation Area Series*, 19: 1-135. (Published, Director, *Zool. Surv. India*, Kolkata).
- Editor-Director. (2005). Fauna of Sambhar Lake (Rajasthan). *Wetland Ecosystem Series*, 6: 1-200. (Published, Director *Zool. Surv. India*, Kolkata).

- Editor-Director. (2009). Faunal Resources of Tal Chhapar Wildlife Sanctuary, Rajasthan. *Conservation Area Series,* 38: 1-66. (Published, Director, *Zool. Surv. India*, Kolkata).
- Fraser. F.E. (1933). The Fauna of British India, including Ceylon & Myanmar, Odonata, Vol. I. Taylor & Francis Ltd., London: 423.
- Fraser, F.E. (1934). The Fauna of British India, including Ceylon & Myanmar, Odonata, Vol. II. Taylor &: Francis Ltd., London: 398.
- Fraser, F.E. (1936). The Fauna of British India, including Ceylon & Myanmar, Odonata, Vol. III. Taylor& Francis LId., London: 461.
- Gaur, Shalini, Kumar, Sanjeev and Rathore, N.S. 2013. Important species of Lizards in the Thar Desert of Rajasthan. In: Faunal heritage of Rajasthan, India: General Background and Ecology of Vertebrates. B. K. Sharma, Seema Kulshreshtha and Asad R. Rahmani (Eds), Springer. Volume I: 287-293.
- Gayan N.C. (1999). A synopsis of the reptiles of Gujarat, western India. *Hamadryad* 24(l): 1-22.
- Ghosh, A.K., Baqri, Q.H. and Prakash, I. (Eds.). (1996). *Faunal Diversity in the Thar Desert: Gaps in Research.* Scientific Publishers, Jodhpur, India. pp. 1-410.
- Ghosh, P.K. (1977). Physiological adaptation of animals to the desert ecosystem. In Desert Ecosystem and its Improvement. (Ed. H.S. Mann).
- Goyal, S.P., Sinha, B., Shah, N. and Panwar, H.S. (1999). Sardar Sarovar Project – a conservation threat to the Indian wild ass (*Equus hemionus khur*). *Biological Conservation,* 88: 277-284.
- Guevara, J. and Sivaperuman, C. (2011). Deserts: Fauna, Flora and Environment. Environmental Science, Engineering and Technology: 1-188.
- Gupta, P.D. (1970). Fauna of Rajasthan. Part 8. Trematoda. *Records of Zoological Survey of India,* Delhi, 62 [1964]: 171-190.
- Gupta, P.D. (1972). Fossil fauna of Rajasthan (India). *Records of Zoological Survey of India,* Delhi. [1968] 66: 153-190.
- Gupta, P.D. (1973). Fauna of Rajasthan, India. Part 3. Mollusca (No. 2). Distribution of some Pelecypoda. *Records of Zoological Survey of India,* Delhi, [1969] 67: 159-163.
- Gupta, P.D. (1975). Fifty years (1920-1969) of faunistic survey of Rajasthan (India). *Records of Zoological Survey of India.* Delhi, [1970] 68: 383-413.
- Gupta, P.D. (1979). Trematode fauna of Rajasthan, India. Part 2. Families Cathaemasiidae, Echinostomatidae and Paramphistomidae. *Rec. Zool. Surv. India,* Delhi, 75 (1-4): 203-208.
- Gupta, R.K. and Prakash, I. (Eds.) (1975). Environmental Analysis of the Thar Desert, viii -f 484 pp., several pls. English Book Depot, Dehra Dun.
- Gupta, S.K. (1985). Handbook of Plant mites of India Z.S.I., Calcutta, 1-520.
- Gupta, S.K. and Gupta, Y.N. (1976). Notes on a collection of tetranychoid mites from Gujarat (Acarina.Trombidiformes). *Acar. Newsl No. 3:* 6-7.
- Hazra, A.K. and Mandal, G.P. (2007). Pictorial Handbook of Collembola (Insecta: Apterygota) (Part -1). Zool. Surv. Ind. Hand book Series.Pages: 1-34 +plates 4.

(Published by the Director, Zool. Surv. India, Kolkata). Himmatsinhji, 1984. On the presence of the *Pangolin Manis crassicaudata* and a fox *Vulpes* sp. in Kutch. *J. Bombay nat. Hist. Soc.*, 81: 686-887.

- Himmatsinhji, (1984). On the presence of the *Pangolin Manis crassicaudata* and a fox *Vulpes* sp. in Kutch. *J. Bombay nat. Hist. Soc.*, 81: 686-887.
- IUCN (2021). The IUCN Redlist of Threatened Species. Version 2020-2. www. iucnredlist.org.
- Jakher, G.R., Dey, T., Misra, D. and Bhargava, S.C. (1981). Diurrnal variations in physico-chemical factors and zooplankton population in Balsamand Lake, Jodhpur. *Geobios*, Jodhpur, 8 (3): 119-122.
- Johal. M.S. and Dhillon, K.S. (1981). Ichthyofauna of Ganganagar district (Rajasthan), India. *Res. Bull. Panjab Univ.* 32: 105-110.
- Johal, M.S., Chauhan, R.C. and Gupta, M.L. (2000). Ichthyofauna of Thar Desert of Rajasthan, India. *Res Bull Punjab Univ*. 50:73-89.
- Joseph, A.N.T. and Parui, P. (1992). Asilidae (Diptera) from Gujarat, India. *Rec. zool. Surv. India*, 92(1-4): 123-129.
- Julka, J.M., Paliwal, R. and Kathireswari. P. (2009). Biodiversity of Indian Earthworms-overview. In Proceedings of Indo-Us Workshop on Vermitechnology in Human Welfare, Edwards, C.A.
- Kankane, P.L. (2000). Status survey of Chinkara and Desert Cat in Rajasthan. *Rec. zool. Surv. India, Occ. Paper*. 179: 1-71.
- Kankane, P.L. and Kumar, Sanjeev, (2016). Gaps in Faunal Exploration and Research in the Thar Desert. In: National Seminar on Faunal Diversity of the Desert Ecosystem: *Lead papers and Abstracts*: 25-33. (Eds.) Kailash Chandra, Sanjeev Kumar & H. S. Banyal.
- Kankane, P.L., Sharma, G. and Venkataraman, K. (2014). Status survey of some mammals in Gujarat with special reference to Chinkara and Desert Cat. 1-302. (published by Director, Zoological Survey of India, Kolkata).
- Kazmi S.I., Bhattacharyya A.K. and Faisal M. (2002). On a collection of Lepidoptera (Butterfly) fauna in the Desert Regional Station, Zoological Survey of India. *Rec. zool Surv. India,* 101 (Part I & II): 1-18.
- Kazmi S.I. and Ramamurthy V.V. (2004). Coleoptera (Insecta) Fauna from the Thar Desert of Rajasthan, India. *Rec. zool Surv. India.*
- Khanna, V. (2001). A checklist of Scolopendrid centipedes from India (Chilopoda: Scolopendromorpha). *Annals of Forestry*, 9(2): 199-219.
- Krishna, D. and Prakash, I. (1960a). A note on mammals of Rajasthan desert, [in Hindi,] *Vijnana Parishad Anusandhan Patrika,* Allahabad, 3 (4): 203-214.
- Kumar, Sanjeev (2016). Avian diversity of Desert Ecosystem of Rajasthan and Gujarat, India. In: National Seminar on Faunal Diversity of the Desert Ecosystem: *Lead papers and Abstracts*: 73. (Eds.) Kailash Chandra, Sanjeev Kumar & H. S. Banyal.
- Kumar, S. (2017). Faunal Diversity of Desert Ecosystem of Rajasthan and Gujarat. Report submitted to Director, Zool. Surv. India, Kolkata. pp. 1-328. (Unpublished)

- Kumar, Sanjeev & Banyal, H.S. (2017). Invasive alien vertebrate species diversity, their impact on biodiversity and status in Rajasthan and Gujarat, India. *ENVIS Newsletter*. 23(1): 16-20.
- Kumar, Sanjeev and Banyal, H.S. (2018). *Faunal Diversity of Kachchh Biosphere Reserve, Gujarat.* (Vertebrates). *Conservation Area Series*, 64: 1-127. (Published by the Director, *Zool. Surv. India*, Kolkata).
- Kumar, Sanjeev, Banyal, H.S., Bano, R. and Raina, R.H. (2020). Locust Plague. *ZSI E-News,* Vol. 15:18-20.
- Kumar S. and Gaur, S. (2009). Herpetofauna, Fauna of Talchhapar Wildlife Sanctuary, Rajasthan, *Conservation area series*, 38: 41-48. (Published by the Director, *Zool. Surv. India*, Kolkata).
- Kumar, Sanjeev and Kumar, S. (2015a). Sighting record of Oriental Honey Buzzard: *Pernis ptilorhynchus* (Temminck, 1821) from Sun city Jodhpur, Rajasthan. *J. Eniv. Bio. -Sci.,* 29 (2): 405-408.
- Kumar, Sanjeev and Kumar, S. (2015b). Baya weaver *Ploceus philippinus* (Linnaeus, 1760) nesting on Bottlebrush tree *Callistemon* in Jodhpur, Rajasthan, India.*J. Bombay Nat. Hist. Soc.,* 112 (1): 35.
- Kumar S., Sivaperuman C., Pardesi M and Baqri Q. H. (2002). Comb Duck, *Sarkioliornis melanotus* (Pennant, 1769) in the Thar Desert of Rajasthan, India. *Newsletters for Birdwatcher*, 42 (5): 102.
- Kumar, A., Kankane, P.L. and Baqri, Q. H. (2006). Geo-spatial Atlas for the wetland Birds of Thar Desert, Rajasthan. Published by the Director, Zool. Surv. India, Kolkata. 202 pp.
- Kumar, R. and Kumar, A. (2009). Studies on Drosophilids (Diptera: Drosophilidae) of Gujarat State in India. *Drosophila Information Service*, 92, p.106.
- Kumar, S., Sharma, S.K., Kant, T. and Emmanuel, C.J.S.K., (2007). Emergence of gall inducing insect *Leptocybe invasa* (Hymenoptera: Eulophidae) in *Eucalyptus* plantations in Gujarat, India. *Indian Forester*, 133(11): 1566-1568.
- Kumawat, R. & Purohit, A. (2020). Snake diversity in golden triangle of Thar Desert of Rajasthan, India. *Uttar Pradesh Journal of Zoology*. **41**(24): 27-32.
- Laity, J.J. 2008. *Desert and Desert Environments*.Wiley-Blackwell Publication.1-360.
- Ishnava, K., Ramarao, V., Mohan, J.S.S. and Kothari, I.L. (2011). Ecologically important and life supporting plants of little Rann of Kachchh, Gujarat. *J. Ecol. Nat. Environment*, 3 (2): 33-38.
- Mahajan, K.K. (1969). Fauna of Rajasthan, India, Part 2. Protozoa (No. 1). Rec. zool Surv. India, 61, [1963]:377-401.
- Mahajan, K.K. (1971). Fauna of Rajasthan, India. Part 10. Protozoa No.2. Records of Zoological Survey of India, Delhi, 63 [1965]: 45-76.
- Mahajan, K.K. (1977). Fauna of Rajasthan Protozoa (No. 3). Rec. zool Surv. India 72: 213-225.
- Mahajan, K.K. (1977). Fauna of Rajasthan, India. Protozoa (No. 3). Rec. zool Surv. India 71 [1973]: 143-148.

- Mahajan, K.K. and Chandra, M. (1976). Report on a collection of leeches from Rajasthan, India. Records of Zoological Survey of India, Delhi, 71 [1973]: 143-148.
- Mahajan, K.K. and Mukherjee, R. (1977). On some rectal Protozoa from frogs and toads of Rajasthan. Newsletter Zoological Survey of India, Calcutta, 3 (6): 365-367.
- Mandal, G. P. (2012). Collembola (Hexapoda) from Thar Desert Rajasthan, India. Nova Science Publishers, Inc. USA. Book chapter, 11:177-183.
- Mandal, G.P. and A.K. Hazra. (2005). Notes on some Collembola (Apterygota: Insecta) from Rajasthan. Rec. Zool. Surv. Ind. 104 (Part 1-2): 1-6.
- Mann, H.S. (1988). Future of the Indian Desert, pp 307-313. In *Desert Ecology.* (Prakash, I. Ed.). Jodhpur: Scientific Publishers.
- Mathur, R.B. and Mathur, S. (1963). Mites associated with stored grains/products in Haryana, India. *Indian J. Acar., 7:*44- 52.
- Mohnot, S.M. (1981). On langurs of Jodhpur. *Abstracts natnl. Sympos. Evaluation of Our Environment* (Jodhpur, Apr. 1981), Jodhpur: 47.
- Mukherjee, A.K. (1995). Birds of arid and semi-arid tracts of northwestern, India. Occ. *paper No. 142, Rec. zool Surv. India.*
- Mukherjee, R.P. (1970). Fauna of Rajasthan, India, Part 9 - Cestoda.*Rec. zool Surv. India,* (1964), 62 (3 & 4): 191 -215.
- Mukherjee, T.K., Ehrmann, R. and Chatterjee, P. (2014). Checklist of Mantodea (Insecta) from India. *Primus Serial Publication of the Centre for Entomological Studies An*kar, 30:1-61.
- Nandi, B.C. (1992). Sarcophagid flies (Diptera: Sarcophagidae) from Gujarat, India. *In Proceedings of the Zoological Society* (Vol. 45, No. 2, pp. 187-196).
- Pal, T.K., Ray, S.K., Talukder, B. and Jaitly, A.K. (2006). On the Megainvertebrate fauna (Mollusca, Branchiopoda, Echinodermata) of Cenozoic and Mesozoic of Kachchh, Gujarat and their stratigraphic implications. *Rec. zool. Surv. India, Occ. Paper* No. 247: 1-46.
- Prakash, I. (1963). Taxonomical and ecological account of the mammals of Rajasthan desert. *Annals of Arid Zone*, Jodhpur, 1: 142-162.
- Prakash, I. (1994). Mammals of the Thar desert. Scientific Publishers.
- Prasad, M. and Thakur, R.K. (1984). On a collection of Odonata from Gujarat State, India. *Fraseria*, No. 6: 24-25.
- Purohit, A.K. and Senacha, K.R. (2002). A review of microchepteran eco-status in Mandore garden, Jodhpur. *Jour. Nature Conservation*, 14: 251-262.
- Purohit, A.K. and Senacha, K.R. (2003). Distribution of chiropteran fauna in and around the Bikaner of Great Indian Desert. *Tiger Paper*, 30(4): 3-8.
- Purohit, A.K. and Senacha, K.R. (2004). Distribution of bats in and around Jaisalmer of the Great Indian Desert, India. *Vespertilio*, 8: 99-104.
- Purohit, A.K., Vyas, K.B. and Senacha, K.R. (2006). Population dynamics of bats in and around Jodhpur of Great Indian Desert. *Tiger Paper*, 33(3): 15-22.
- Senacha, K.R., Vyas, K.B. and Purohit, A.K. (2006). New records of Short-nosed Fruit Bat *Cynopterus sphinx* (Vahl, 1797) from Thar Desert, Rajasthan. *Zoos'Print Journal*, 21(10): 2419-2420.

- Rathore, N.S. and Bhattacharyya, A.K. (2004). *Termite (Insecta: Isoptera) Fauna of Gujarat and Rajasthan: Present State of Knowledge. Occ. Pap. No*. 223. Records Zoological Survey of India, pp. 1-77.
- Ray, H.C. and Mukherjee, A. (1969). Fauna of Rajasthan, India. Pt. 3. Mollusca. *Records of Zoological Survey of India,* Delhi, [1963]: 61 403-436.
- Rodger, W.A., Panwar, H.S. and Mathur, V.B. (2002). Wildlife Protected Area Network in India: A review, Executive Summary. Wildlife Institute of India, Dehradun. pp. 1-51.
- Roonwal, M.L. (1983). Fauna of the Great Indian Desert. In: *Desert resources and technology*. (Ed.) A.N. Singh. Scientific Publisher; Geotech-Academia; p 1-86.
- Roonwal, M.L. and Bose, G. (1969). Fauna of Rajasthan, India. Part 4. A checklist of Rajasthan termites (Insecta: Isoptera). Records of Zoological Survey of India, Delhi. 61 [1963]: 437-450, 4 pls.
- Saha, U., Ghosh, M. and Pal, T.K. (2004). Animal remains excavated from Lothal archaeological site (Gujarat) and relevance of the fauna in this ancient civilization. *Rec. zool. Surv. India, Occ. Paper No*. 222: 1-162.
- Saxena, M.M. (1996). Freshwater sponges in the Thar Desert. In: *Faunal Diversity in the Thar Desert: Gaps in Research*. (Eds.) A.K. Ghosh, Baqri, Q.H. and I. Prakash. 37-41. Scientific Publishers, Jodhpur.
- Saxena, M.M. (2017). Porifera (Sponge). In: *Current Status of Freshwater Faunal Diversity of India*: 55-66 (published by the Director, *Zool. Surv. India*, Kolkata).
- Senacha, K.R. (2003). Eco-status and demographic changes among the chiropterans of the Thar desert with special reference to Jodhpur. Ph.D. Thesis, JNV University, Jodhpur, India. pp. 175.
- Sewak, R. (2004). Insecta: Coleoptera: Scarabaeidae: Coprinae, In: *State Fauna Series No.8, Gujarat* (*Part 2*), pp. 105-125. Editor-Director; Zoological Survey of India, Kolkata.
- Sewak, R. (2009a). Dung Beetles (Insecta: Coleoptera: Scarabaeidae: Coprinae) of Thar Desert of Gujarat. *Rec. zool. Surv. India. Occ. Paper*, 295: 1-48.
- Sewak, R. (2009b). Dung Beetles (Insecta: Coleoptera: Scarabaeidae: Coprinae) of Rajasthan. *Rec. zool. Surv. India. Occ. Paper*, 296: 1-106.
- Shah, G.L. (1978). Flora of Gujarat state, part I & II. Sardar Patel University, Vallabh Vidhyanagar (Gujarat).
- Sharma, A., Lata, P., Rathore, N.S. and Thakur, R. (2013). A study on variations in population density of gastropods in a village pond near Bikaner, Rajasthan. *J. Exp. Biol. Agri. Sci.*, 1 (3): 180-185.
- Sharma Manju, Srivastava Meera. (2010). Lepidopteran fauna of an agro-ecosystem in Western Rajasthan: A short-term surveillance. *Journal of Entomological Research.* 34 (3): 249-258.
- Sharma, G. (2009). Studies and status of damselflies and dragonflies (Odonata: Insecta) of Arid and Semi-Arid regions of India. *Hexapoda*, 16 (1): 36-39.
- Sharma, G. (2013). A review on the Studies on Faunal diversity, status, Threats and Conservation of Thar Desert or Great Indian Desert Ecosystem. *Biological Forum– An International Journal.* 5(2):81-90.

- Sharma, G. (2015). Pictorial Handbook on Damselflies and Dragonflies (Odonta: Insecta) of Rajasthan: 1-266, + 93 photoplates. Publ.: Director, ZSI, Kolkata.
- Sharma, G. (2018). Studies on Odonata and Lepidoptera Fauna of Desert Ecosystem of Rajasthan. *Rec. zool. Surv. India*, Occ. Paper No., 394: 1-90.
- Sharma, G. and Dhadeech, S.N. (2010). Studies on Damselflies and Dragonflies (Odonta: Insecta) in and around Thar Desert at Osian, Jodhpur, Rajasthan, India. *Proceedings of the National Seminar on Impact of Climate Change on Biodiversity and Challenges in Thar Desert*: 195-197. Eds.: Ramakrishna, K. Chandra, P. Bohra & G. Sharma.
- Sharma, G. and Kankane, P.L. (2012). Fauna of Ecosystems of India - Thar Desert. 1-18. Publ.: Director, ZSI, Kolkata.
- Sharma, R.C. (1981). *Hemidactylus porbandarensis* a new geckonid lizard from Gujarat. *India. Bull. Zool. Surv. India,* 4 (1): 1-2.
- Sharma, R.C. (1982). Taxonomic and ecological studies on the reptiles of Gujarat. *Records of the Zoological Survey of India,* 80: 85-108.
- Sharma, Rashmi. (2015). Faunal Diversity of Insects of Fresh Water Lake of Ajmer, Rajasthan. OSR Journal of Pharmacy and Biological Sciences (IOSR-JPBS) e-ISSN: 2278-3008, p-ISSN: 2319-7676. Volume 10, Issue 6 Ver. IV (Nov-Dec. 2015), PP 39-43. www.iosrjournals.org.
- Sinha, Y P. (1980). The bats of Rajasthan: Taxonomy and zoogeography. *Records of Zoological Survey of India,* Delhi, 76: 7-63.
- Sirawat, S. (2016). Importance of Anuran Diversity in the Thar Desert of the Rajasthan state. In: National Seminar on Faunal Diversity of the Desert Ecosystem: *Lead papers and Abstracts*: 94. Eds. Kailash Chandra, Sanjeev Kumar & H. S. Banyal.
- Sivaperuman, C., Baqri, Q.H., Ramaswamy, G. and Naseema, M. (2009). *Faunal Ecology and Conservation of the Great Indian Deser*t, Springer Publications, 1-222.
- Soota, T.D. and Saxena, M.M. (1983). Sponge fauna of some waters of Rajasthan and its ecology. *Trans. Isdt. Ucds*., 8 (2): 131 -133.
- Stoliczka, F. (1872a). Notes on the reptilian and amphibian fauna of Kutchh (Cutch) of *Proceedings of the Asiatic Society Bengal* May: 71-85.
- Stoliczka, F. (1872b). Notices of the mammals and birds inhabiting Kachh. *J. Asiat. Soc. Beng*., 41: (Part II): 211 258.
- Tak N. and Rathore N. S. (1996). Ant (Formicidae) Fauna of Thar Desert: Faunal Diversity in The Thar Desert: Gaps in Research (Edited by Ghosh, Baqri and Prakash). Scientitic Publishers. pp. 271-279.
- Tak, N. (2009). Ants (Hymenoptera: Formicidae) of the Thar Desert of Rajasthan and Gujarat. *In Faunal Ecology and Conservation of the Great Indian Desert* (pp. 41-48). Springer Berlin Heidelberg.
- Tak, N. and Rathore, N.S. (2004). Insecta: Hymenoptera. In: *Fauna of Desert National Park Conservation Area Series*, 19. *Zool. Surv. India*. pp. 85-90.
- Thakur, M. L. (1981). Revision of Termite Genus Odontotermes Holmgren from India (Isoptera :Termitidae : Macrotermitinae). *Indian For. Rec*., 14 (2): 1-183.

- Thakur, R. K. (1989). Termite fauna of Gujarat (Insecta: Isoptera); *Indian For. Rec.* (N.S.), Ent., 15 (1): 1-76.
- Tripathi, G., Deora, R. and Sharma, B.M. (1997). Community and habitat structure of termites and centipedes of arid land. *J. Exep. Zool.,* 10 (2): 291-300.
- Tripathi, Y.C. and Arya, R. (2002). Floral Resources of Indian Thar Desert. In: *Bioresource Technology* (Eds.) G. Tripathi 18-33, CBS Publishers and Distributors, New Delhi.
- Tyagi, B.K. and Baqri, Q.H. (2005). *Changing Faunal Ecology in the Thar Desert.* Scientific Publishers, Jodhpur, 1: 367.
- Tyagi, K. and Kumar, V. (2016). Thrips (Insecta: Thysanoptera) of India- An Updated checklist. *Halteres*, 7: 64-98.
- Varshney, R.K. (1992). A check list of the scale insects and mealybugs of South Asia. Part I. *Rec. zool.Surv. India, Occ. Paper*, No. 139: 1-152.
- Varshney, R.K. (2002). A check list of the scale insects and mealybugs of South Asia. Part 2. *Rec. zool. Surv. India, Occ. Paper*, No. 191: 1-147.
- Vazirani, T.G. (1973). Contribution to the study of Aquatic beetles (Coleoptera) 12 on a collectionof Dytiscidae from Gujarat with discription of new species. *Rec. zool. Surv. India*. 67: 287-302.
- Vyas, R. (1998). The reptiles of Gujarat State: updated distribution. *Tiger Paper,* 25 (l): 8-14.

CHAPTER 5

Bats of Rajasthan—An Updated Review

C. Srinivasulu
Wildlife Biology & Taxonomy Lab, Department of Zoology, University College of Science, Osmania University, Hyderabad-500 007, Telangana State, India and Centre for Biodiversity & Conservation Studies, Osmania University, Hyderabad-500 007, Telangana State, India

ABSTRACT

Rajasthan's bat diversity contains 26 species from 19 genera and 8 families, which includes three species of fruit bats and 23 species of insectivorous bats. *Eptesicus serotinus*, *Barbastella darjelingensis* and *Myotis blythi* are known based on historic records, but their current status and range in Rajasthan are unknown. Only eleven species of bats are widespread in their distribution in Rajasthan and have been frequently observed in the Aravalli Hills, non-desert areas and fringes of the Thar desert. Greater Mouse-tailed Bat *Rhinopoma microphyllum* is the only species that was observed in the deeper regions of the Thar desert. Owing to changes in vegetation pattern due to the Indira Gandhi Canal the fruit bats are spreading in to desert regions too.

Key words: Chiroptera, taxonomy, distribution, status, India

INTRODUCTION

Bats have been increasingly becoming a group of animals that are important to humankind, not only due to their unique ability of sustained flight but also due their natural ability to be the natural reservoirs to many pathogens including virus. These remarkable capabilities are due to a long evolutionary history including morphological and physiological adaptations modifying over time. Representing nearly a quarter of the known mammals, bats are found throughout the world excepting the cold Arctic, the Antarctica and a few Oceanic Islands (Wilson & Mittermeier, 2019). Order Chiroptera includes about 1435 species of bats world over in rather two unequal suborders – the Yinpterochiroptera (consisting 419 species belonging to family Craseonycteridae, Hipposideridae, Megadermatidae, Pteropodidae, Rhinolophidae, Rhinonycteridae, and Rhinopomatidae) and the Yangochiroptera (consisting 1016 species belonging to family Cistugidae, Emballonuridae, Furipteridae, Miniopteridae, Molossidae, Mormoopidae, Mystacinidae, Myzopodidae, Natalidae, Noctilionidae, Nycteridae, Phyllostomidae, Thyropteridae, and Vespertilionidae) (Simmons and Cirranello, 2020). As many as 133 species of bats are known from India (Srinivasulu *et al.*, 2021).

Srinivasulu *et al.* (2013) compiled a summary of Rajasthan's chiropteran variety based on historical and present primary material, as well as surveys and investigations conducted around the state. Since then a few more studies have been published, and this review an update is being provided on the bats of Rajasthan to the best of our present knowledge (cutoff date: 30 September, 2021). The reader is urged to refer to Srinivasulu *et al.* (2013) for further details on sources of information. New data since the publication of the last consolidated works has been compiled from various sources, and the references cited here include only the literature cited in this chapter.

Bats of Rajasthan

The bat diversity of Rajasthan is represented by 26 species (Table 1), of which a few species, like *Eptesicus serotinus pachyomus*, *Barbastella darjelingensis* and *Myotis blythi* are of doubtful records or might have had their distribution in the historic times, but not now. These species have not been reported in the recent past by any researchers in Rajasthan. As many as 18 species are known to occurring in the Thar desert region (Sinha, 1996), and of these four to five species have been reported not being sighted in the Thar since late 1970s (Dookia *et al.*, 2017).

Table 1. *Bats (Order Chiroptera) of Rajasthan, India*

S. No.	Family	Species
Suborder Yinpterochiroptera		
1.	Pteropodidae	*Pteropus medius*
2.		*Rousettus leschenaultii*
3.		*Cynopterus sphinx*
4.	Rhinopomatidae	*Rhinopoma hardwickii*
5.		*Rhinopoma microphyllum*
6.	Megadermatidae	*Lyroderma lyra*
7.	Rhinolophidae	*Rhinolophus Lepidus*
8.	Hipposideridae	*Asellia tridens*
9.		*Hipposideros fulvus*
10.		*Hipposideros lankadiva*
Suborder Yangochiroptera		
11.	Emballonuridae	*Taphozous perforatus*
12.		*Taphozous longimanus*
13.		*Taphozous nudiventris*
14.		*Taphozous melanopogon*
15.	Molossidae	*Tadarida aegyptiaca*
16.		*Chaerephon plicatus*
17.	Vespertilionidae	*Eptesicus serotinus*
18.		*Hesperoptenus tickelli*
19.		*Barbastella darjelingensis*
20.		*Scotophilus kuhlii*

21.		*Scotophilus heathii*
22.		*Pipistrellus tenuis*
23.		*Pipistrellus ceylonicus*
24.		*Scotozous dormeri*
25.		*Myotis blythi*
26.		*Kerivoula picta*

Order Chiroptera Blumenbach, 1779

Suborder Yinpterochiroptera Springer, Teeling, Madsen, Stanhope and Jong, 2001

Family Pteropodidae Gray, 1821

Genus *Pteropus* Brisson, 1762 [Flying Foxes]

1. **Indian Flying Fox *Pteropus medius* Temminck, 1825**

   Monogr. Mamm. 1: 176.

   Distribution in Rajasthan: *Pteropus medius* Temminck, 1825 has been reported from most of Rajasthan including desert area (Prakash, 1960, 1961; Sinha, 1980; Gaur, 1981; Advani, 1982a,b; Purohit & Senacha, 2002, 2003; Senacha, 2003, 2006; Tak & Dookia, 2003; Dookia & Tak, 2004; Purohit & Vyas, 2006; Purohit, Vyas & Senacha, 2006; Srinivasulu & Srinivasulu 2006).

   Genus *Rousettus* Gray, 1821 [Rousettes]

2. **Leschenault's Rousette *Rousettus leschenaultii* (Desmarest, 1820)**

   Mammalogie, in Encyclop. Méthod. 1: 110.

   Distribution in Rajasthan: *Rousettus leschenaultii leschenaultii* (Desmarest, 1820) has been reported mostly from Aravalli Hills and non-desert regions, and from a few locations in the Thar desert of Rajasthan (Andersen, 1912; Wroughton, 1918; Sinha, 1980; Gaur, 1981; Bates *et al.*, 1994a; Bates & Harrison, 1997; Purohit *et al.*, 2006; Srinivasulu & Srinivasulu, 2006).

   Genus *Cynopterus* Cuvier, F., 1824 [Short-nosed Fruit Bats]

3. **Greater Short-nosed Fruit Bat *Cynopterus sphinx* (Vahl, 1797)**

   Skr. Nat. Selsk. Copenhagen 4(1): 123.

   Distribution in Rajasthan: *Cynopterus sphinx* (Vahl, 1797) has been reported mostly from Aravalli Hills and non-desert regions, and from a few locations in the Thar desert of Rajasthan (Sinha, 1980; Gaur, 1981; Advani, 1982a,b; Purohit & Senacha, 2002; Senacha, 2003; Dookia, 2004; Purohit & Vyas, 2006; Purohit *et al.*, 2006; Senacha *et al.*, 2006; Srinivasulu & Srinivasulu, 2006).

   Family Rhinopomatidae Bonaparte, 1838

   Genus *Rhinopoma* E. Geoffroy, 1818 [Mouse-tailed Bats]

4. **Lesser Mouse-tailed Bat *Rhinopoma hardwickii* Gray, 1831**

   Zool. Misc., 1: 37.

   Distribution in Rajasthan: *Rhinopoma hardwickii* Gray, 1831 has been reported mostly from Aravalli Hills and non-desert regions, and from a few locations in the

Thar desert of Rajasthan (Sinha, 1980; Gaur, 1981; Bates & Harrison, 1997; Purohit & Senacha, 2002, 2003; Senacha, 2003; Senacha & Purohit, 2005; Purohit & Vyas, 2006; Purohit *et al.*, 2006; Srinivasulu & Srinivasulu, 2006).

5. **Greater Mouse-tailed Bat *Rhinopoma microphyllum* (Brünnich, 1872)**

   Dyrenes Historie, 1: 50, pl. 6, fig.1-4.

   Distribution in Rajasthan: *Rhinopoma microphyllum kinneari* Wroughton, 1912 has been reported mostly from Aravalli Hills and non-desert regions, and from a few locations in the Thar desert of Rajasthan (Prakash, 1960; Sinha, 1980; Gaur, 1981; Advani, 1981, 1982a,b; Agarwal & Gupta, 1982; Lall, 1985; Purohit & Senacha, 2001, 2002, 2003; Senacha, 2002, 2003; Purohit & Vyas, 2006; Purohit *et al.*, 2002, 2006; Senacha *et al.*, 2006; Srinivasulu & Srinivasulu, 2006).

   Family Megadermatidae H. Allen, 1864

   Genus *Lyroderma* Peters, 1873 [Greater False Vampire Bats]

6. **Greater False Vampire Bat *Lyroderma lyra* (E. Geoffroy, 1810)**

   Ann. Mus. Hist. Nat. Paris, 15: 190.

   Distribution in Rajasthan: *Lyroderma lyra* (E. Geoffroy, 1810) has been reported mostly from Aravalli Hills and non-desert regions, and from a few locations in the Thar desert of Rajasthan (Prakash, 1960; Ramaswami & Anand Kumar, 1963; Gaur, 1979, 1981; Sinha, 1980; Advani, 1981a, 1982b; Gupta & Trivedi, 1989; Srinivasulu & Srinivasulu, 2006). Purohit & Senacha (2002; 2004a,b) and Purohit *et al.* (2006) reported its absence in Jodhpur and its vicinity.

   Family Rhinolophidae Bell, 1836

   Genus *Rhinolophus* Lacépède, 1799 [Horseshoe Bats]

7. **Blyth's Horseshoe Bat *Rhinolophus lepidus* Blyth, 1844**

   J. Asiat. Soc. Bengal, 13: 486.

   Distribution in Rajasthan: *Rhinolophus lepidus* Blyth, 1844 has been reported mostly from Aravalli Hills and non-desert regions, and from a few locations in the Thar desert of Rajasthan (Prakash, 1961; Sinha, 1980; Gaur, 1981; Purohit & Senacha, 2002, 2003; Senacha, 2003; Purohit & Vyas, 2006; Purohit *et al.*, 2006).

   Family Hipposideridae Lydekker, 1891

   Genus *Asellia* Gray, 1838 [Trident Bats]

8. **Geoffroy's Trident Leaf-nosed Bat *Asellia tridens* (E. Geoffroy, 1813)**

   Ann. Mus. Natn. Hist. Nat. Paris 20: 265.

   Distribution in Rajasthan: *Asellia tridens* (E. Geoffroy, 1813) has been reported only from Gajroop Sagar in Jaisalmer District (Senacha & Dookia, 2013).

   Genus *Hipposideros* Gray, 1831 [Leaf-nosed Bats]

9. **Fulvus Leaf-nosed Bat *Hipposideros fulvus* Gray, 1838**

   Mag. Zool. Bot., 2: 492.

   Distribution in Rajasthan: *Hipposideros fulvus pallidus* Andersen, 1918 has been reported mostly from Aravalli Hills and non-desert regions, and from a few locations in the Thar desert of Rajasthan (Sinha, 1980; Gaur, 1981; Bhupathy, 1987; Bates *et al.*, 1994b; Bates & Harrison, 1997). Purohit & Senacha (2002; 2004a) and Purohit *et al.* (2006) report this species being absent from Jodhpur and its vicinity; however, it has been recently reported from Daijar village, 25 km north of Jodhpur (Dookia *et al.*, 2017).

10. Indian Leaf-nosed Bat *Hipposideros lankadiva* Kelaart, 1850

J. Sri Lanka Branch Asiat. Soc., 2(2): 216.

Distribution in Rajasthan: *Hipposideros lankadiva indus* (Andersen, 1918) has been reported only from Bhim Bharak in Jodhpur District (Wason, 1978). No recent sighting reports from Rajasthan.

Suborder Yangochiroptera Springer, Teeling, Madsen, Stanhope and Jong, 2001

Family Emballonuridae Gervais, 1855

Genus *Taphozous* E. Geoffroy, 1818 [Tomb Bats]

11. Egyptian Tomb Bat *Taphozous perforatus* E. Geoffroy, 1818

Descrip. de L'Egypte 2: 126.

Distribution in Rajasthan: *Taphozous perforatus* E. Geoffroy, 1818 has been historically reported from Jodhpur, Barmer, Dungarpur, Jaisalmer, & Jhalawar (Prakash, 1960; Sinha, 1980; Advani, 1982b). There are no recent records of this species in and around Jaisalmer (Purohit & Senacha, 2002).

12. Long-winged Tomb Bat *Taphozous longimanus* Hardwicke, 1825

Trans. Linn. Soc. Lond. 14: 525.

Distribution in Rajasthan: *Taphozous longimanus* Hardwicke, 1825 has been reported historically from Nasirabad and Kota (Sinha, 1980).

13. Naked-rumped Tomb Bat *Taphozous nudiventris* Cretzschmar, 1830

In Rüppell, Atlas Reise Nördl. Afr., Zool. Säugeth. p. 70.

Distribution in Rajasthan: *Taphozous nudiventris kachhensis* Dobson, 1872 has been reported mostly from Aravalli Hills and non-desert regions, and from a few locations in the Thar desert of Rajasthan (Advani, 1980; Sinha, 1980; Gaur, 1981; Purohit & Senacha, 2002; Senacha, 2003; Purohit & Vyas, 2006; Purohit *et al.*, 2006; Srinivasulu & Srinivasulu, 2006).

14. Black-bearded Tomb Bat *Taphozous melanopogon* Temminck, 1841

Monogr. Mamm. 2: 287.

Distribution in Rajasthan: *Taphozous melanopogon* Temminck, 1841 has been reported only from Udaipur in Rajasthan (Sinha, 1981).

Family Molossidae Gill, 1872

Genus *Tadarida* Rafinesque, 1814 [Free-tailed Bats]

15. Egyptian Free-tailed Bat *Tadarida aegyptiaca* (E. Geoffroy, 1818)

Discrip. De L'Egypte 2: 128, pl. 2, No. 2.

Distribution in Rajasthan: *Tadarida aegyptiaca thomasi* (Wroughton, 1919) has been reported from Aravalli Hills and non-desert regions, and from a few locations in the Thar desert of Rajasthan (Ryley, 1914; Sinha, 1980; Advani, 1982b,c). Surveys in and around Jodhpur did not report its presence (Purohit & Senacha, 2002, 2004a,b; Purohit *et al.*, 2006).

Genus *Chaerephon* Dobson, 1874 [Mastiff Bats]

16. Wrinkle-lipped Free-tailed Bat *Chaerephon plicatus* (Buchanan, 1800)

Trans. Linn. Soc. London, 5: 261, pl.13.

Distribution in Rajasthan: *Chaerephon plicatus* has been reported only from Mt. Abu in Sirohi District (Sinha, 1980). No recent sighting reports from Rajasthan.

Family Vespertilionidae Gray, 1821

Genus *Eptesicus* Rafinesque, 1820 [Serotines]

17. Serotine *Eptesicus serotinus* Schreber, 1774

Die Saugethiere, 1: 167.

Distribution in Rajasthan: *Eptesicus serotinus pachyomus* has collected from Nasirabad in Ajmer District (Sinha, 1980), which is thought to be the type locality of this taxon, which was historically assigned as 'Rajputana' (present day Rajasthan), India. No recent sighting reports from Rajasthan.

Genus *Hesperoptenus* Peters, 1868 [False Serotines]

18. Tickell's Bat *Hesperoptenus tickelli* (Blyth, 1851)

J. Asiat. Soc. Bengal, 20: 157.

Distribution in Rajasthan: *Hesperoptenus tickelli* (Blyth, 1851) has historically been reported only from Nasirabad in Ajmer District (Blanford, 1891; Sinha, 1980). No recent sighting reports from Rajasthan.

Genus *Barbastella* Gray, 1821 [Barbastelle]

19. Darjeeling Barbastelle *Barbastella darjelingensis* Hodgson [in Horsfield], 1855.

Ann. Mag. Nat. Hist. ser. 2, 16: 102.

Distribution in Rajasthan: *Barbastella darjelingensis* (Hodgson, 1855 *in* Horsfield, 1855) has historically been reported only from Nasirabad in Ajmer District (Wroughton, 1918; Ellerman & Morrison-Scott, 1951; Sinha, 1980). No recent sighting reports from Rajasthan.

Genus *Scotophilus* Leach, 1821 [Yellow House Bats]

20. Lesser Asiatic Yellow House Bat *Scotophilus kuhlii* Leach, 1821

Trans. Linn. Soc. Lond., 13: 71.

Distribution in Rajasthan: *Scotophilus kuhlii* Leach, 1821 has been reported from Aravalli Hills and non-desert regions, and from a few locations in the Thar desert of Rajasthan (Sinha, 1980; Advani, 1982b).

21. Greater Asiatic Yellow House Bat *Scotophilus heathii* (Horsfield, 1831)

Proc. Zool. Soc. Lond., p.113.

Distribution in Rajasthan: *Scotophilus heathii* (Horsfield, 1831) has been reported from Aravalli Hills and non-desert regions, and from a few locations in the Thar desert of Rajasthan (Prakash, 1960; Sinha, 1980; Gaur, 1981; Advani, 1982b; Purohit & Senacha, 2002; Senacha, 2003; Purohit & Vyas, 2006; Purohit *et al.*, 2006).

Genus *Pipistrellus* Kaup, 1829 [Pipistrelles]

22. Least Pipistrelle *Pipistrellus tenuis* (Temminck, 1840)

Monogr. Mamm., 2: 229.

Distribution in Rajasthan: *Pipistrellus tenuis mimus* Wroughton, 1899 has been reported from Aravalli Hills and non-desert regions, and from a few locations in the Thar desert of Rajasthan (Prakash, 1960; Sinha, 1980; Advani, 1981b, 1982b; Gaur, 1981; Purohit & Senacha, 2002, 2003; Purohit & Vyas, 2006; Purohit *et al.*, 2006).

23. Kelaart's Pipistrelle *Pipistrellus ceylonicus* (Kelaart, 1852)

Prodr. Faun. Zeylanica, p. 22.

Distribution in Rajasthan: *Pipistrellus ceylonicus indicus* has historically been reported from Mt. Abu in Sirohi District (Ryley, 1914). No recent report its presence in Rajasthan.

Genus *Scotozous* Dobson, 1875 [Dormer's Pipistrelle]

24. Dormer's Pipistrelle *Scotozous dormeri* Dobson, 1875

Proc. Zool. Soc. Lond., p. 373.

Distribution in Rajasthan: *Scotozous dormeri* Dobson, 1875 has been reported from Aravalli Hills and non-desert regions, and from a few locations in the Thar desert of Rajasthan (Sinha, 1980; Advani, 1981c, 1982b; Purohit *et al.*, 2006).

Genus *Myotis* Kaup, 1829 [Mouse-eared Bats]

25. Lesser Mouse-eared Myotis *Myotis blythi* (Tomes, 1857)

Proc. Zool. Soc. Lond., p. 53.

Distribution in Rajasthan: *Myotis blythi* has historically been reported from Nasirabad in Ajmer District (Sinha, 1980). However, Topal (1971) suggests that the type locality could be Nassenabad which is located in the Himalayas. No recent report its presence in Rajasthan.

Genus *Kerivoula* Gray, 1842 [Woolly Bats]

26. Painted Woolly Bat *Kerivoula picta* (Pallas, 1767)

Spicil. Zool., 3: 7.

Distribution in Rajasthan: *Kerivoula picta* has been reported only from Kishore Pura in Alwar District (Sharma, 1986).

CONSERVATION STATUS

All of the 26 bat species found in Rajasthan have a fairly wide worldwide range and are reasonably stable, hence according to IUCN Red List of Threatened Species all belong to Least Concern category. However, one bat species, the Indian Flying Fox, is included in Appendix II of the Convention on International Trade in Endangered Species (CITES).

Note on Distribution

Bats in Rajasthan have been observed largely around towns and villages where vegetation growth is comparatively better, avoiding the extremely parched portions of the Thar desert. The presence of the Indira Gandhi Canal, which influences groundwater availability and vegetation, is causing more and more regions of the Thar desert to turn green. A few species, like *Eptesicus serotinus*, *Barbastella darjelingensis* and *Myotis blythi*, are known only from historical observations or collections, while a few, like *Asellia tridens* and *Kerivoula picta*, have been recorded in the last 25 years. The other species are more common in the Aravalli Hills and non-desert regions of Rajasthan.

Acknowledgements

I am thankful to bat researchers of Rajasthan – Prof. A.K. Purohit, Dr K. R. Senacha, Dr Sumit Dookia and their research teams, and Dr D. L. Bohra, for fruitful discussions on bats since very long. I am thankful to Dr. Bhargavi and Mr. Tariq Ahmed Shah for updating distribution records and conducting field surveys in Rajasthan. I take this opportunity to also express my gratitude to Dr Y. P. Sinha for his guidance and in-depth studies on bats in Rajasthan. I acknowledge encouragement and support extended to my research by the Vice Chancellor, Osmania University and the Head, Department of Zoology, Osmania University, Hyderabad. Last but not the least, I dedicate this chapter in fond memory of my dear friend and colleague – Dr. Parmeshwar S. Chauhan of College of Horticulture & Forestry, Jhalawar, whom we lost in May 2021 to COVID-19 pandemic.

REFERENCES

- Advani, R. (1980). Observations on feeding ecology and behaviour of the Kutch sheath tailed bat, *Taphozous kachhensis kachhensis* in Rajasthan. *Zeitschrift für angewandte Zoologie*, 67(3): 279–285.
- Advani, R. (1981a). Seasonal fluctuation in the feeding ecology of the Indian false vampire, *Megaderma lyra lyra* (Chiroptera: Megadermatidae) in Rajasthan. *Zeitschrift für Säugetierkunde*, 46(2): 90–93.
- Advani, R. (1981b). Some observations on the feeding behaviour of the Indian Pygmy Pipistrelle, *Pipistrellus mimus mimus* Wroughton, 1899 (Mammalia: Chiroptera: Vespertilionitidae) in Rajasthan Desert. *Säugetierkundliche Mitteilungen*, 29(4): 10–12.
- Advani, R. (1981c). Feeding ecology and behaviour of the Dormer's bat, *Pipistrellus dormeri dormeri* (Dobson, 1875), in the India Desert. *Säugetierkundliche Mitteilungen*, 29(4): 56–58.
- Advani, R. (1982a). Feeding, foraging and roosting behaviour of the fruit-eating bats and damage to fruit crops in Rajasthan and Gujarat. *Säugetierkundliche Mitteilungen*, 30(1): 46–48.
- Advani, R. (1982b). Distribution and status of chiroptera species in Rajasthan, India. *Säugetierkundliche Mitteilungen*, 30(1): 49–52.
- Advani, R. (1982c). Feeding ecology of *Tadarida aegyptiaca thomasi* in the Indian desert. *Zeitschrift für Säugetierkunde*, 47: 18–22.
- Advani, R. (1981d). Food and feeding ecology of the rat tailed bat in the Rajasthan Desert. *Acta Theriologica*, 26(8-15): 269–272.
- Agarwal, A, and B.B. Gupta (1982). The gastric morphology and histology of an insectivorous bat – *Rhinopoma kinneari*. *Lynx*, 21: 5–14.
- Andersen, K. (1912). *Catalogue of Chiroptera in the collection of the British Museum. Vol. 1. Megachiroptera*. British Museum of Natural History, London. 854 pp.
- Andersen, K. (1918). *Hipposideros fulvus pallidus*, *Hipposideros lankadiva indus*. *Annals and Magazine of Natural History*, 9th Series 2: 381–382.
- Bates, P.J.J., D.L. Harrison and M. Muni (1994a). The Bats of Western India revisited. Part 1. *Journal of the Bombay Natural History Society*, 91(1): 1–15.

- Bates, P.J.J., D.L. Harrison and M. Muni (1994b). The Bats of Western India revisited. Part 3. *Journal of the Bombay Natural History Society,* 91(3): 360–380.
- Bates, P.J.J. and D.L. Harrison (1997). *Bats of the Indian Subcontinent.* Harrison Zoological Museum, Sevenoaks, UK. p xvi + 258.
- Bhupathy, S. (1987). Occurrence of the bicoloured leaf-nosed bat (*Hipposideros fulvus*) in Rajasthan. *Journal of the Bombay Natural History Society*, 84(1): 199–200.
- Blanford, W.T. (1891). The Fauna of British India, *Mammalia. Part 2. Taylor & Francis,* London.
- Dookia, S. and J.R. Tak (2004). Status and distribution of Indian Flying Fox (*Pteropus giganteus* Brunnich) in Thar Desert of Rajasthan. *BatNet - CCINSA Newsletter,* 5(1): 7–8.
- Dookia, S. (2004). Occurrence of the Short-nosed Fruit Bat (*Cynopterus sphinx* Vahl, 1797) in Thar Desert of Rajasthan. *Zoos'Print Journal*, 19(9): 1629.
- Dookia, S., G. Singh and R. Mishra (2017). Re-sighting record of Fulvous Leaf-nosed Bat *Hipposideros fulvus* Gray, 1838 (Mammalia: Chiroptera: Hipposideridae) from Thar Desert, Rajasthan, India. *Journal of Threatened Taxa*, 9(1): 9764–9767.
- Ellerman, J.R. and T.C.S. Morrison-Scott (1951). Checklist of Palearctic and Indian Mammals – 1758 to 1946. British Museum (Natural History), London. p. 810.
- Gaur, B.S. (1979). Vanishing of Indian False Vampire *Megaderma lyra lyra* (Geoffroy) from Jodhpur. *Comparative Physiology & Ecology*, 4(4): 260.
- Gaur, B.S. (1981). Ecology of bats of the Indian Desert. Unpublished Ph.D. Thesis, Jodhpur University, Jodhpur, India. 110 pp.
- Gupta, S.K.and K.K. Trivedi (1989). Nematode parasites of vertebrates. On two new species of the genus *Litomosoides* Chandler, 1931 (Family: Dipetalonematidae Wehr, 1935) from microbats of Udaipur, Rajasthan, India. *Indian Journal of Helminthology*, 41(suppl.): 152–161.
- Lall, S.B. (1985). Folliculogenesis in *Rhinopoma kinneari* Wroughton (Microchiroptera: Mammalia). *Myotis*, 23&24: 37–44.
- Prakash, I. (1960). Breeding mammals in Rajasthan Desert, India. *Journal of Mammalogy*, 41(3): 386–389.
- Prakash, I. (1961). Die Saugetiere der Rajasthan – Wustein Indien. *Natur und Volk*, 91: 443–448.
- Purohit, A.K. and K.R. Senacha (2002). A review of microchepteran eco-status in Mandore garden, Jodhpur. *Journal of Nature Conservation*, 14: 251–262.
- Purohit, A.K., B.S. Gaur, K.R. Senacha, V.K. Vijayakrishna and A.K. Chhangani (2002). Rare observation of puppies preying upon a greater mouse-tailed bat: *Rhinopoma microphyllum kinneari*. *Flora and Fauna*, 8(2): 115–116.
- Purohit, A.K. and K.R. Senacha (2002). Present population status of *Rhinopoma microphyllum kinneari* at J.N.V. University, Jodhpur. *BatNet - CCINSA Newsletter*, 2(1): 16.
- Purohit, A.K. and K.R. Senacha (2004a). Demographic changes among bats in and around Jaisalmer of Great Indian Desert. *Cheetal*, 42(1-2): 25–34.
- Purohit, A.K.and K.R. Senacha (2004b). Distribution of bats in and around Jaisalmer of the Great Indian Desert, India. *Vespertilio*, 8: 99–104.

- Purohit, A.K. and K.R. Senacha (2003a). Electrocution of Indian Flying Foxes, *Pteropus giganteus giganteus*, in Jodhpur, India. *Bat Research News*, 44(2): 63–64.
- Purohit, A.K. and K.R. Senacha (2003b). Distribution of chiropteran fauna in and around the Bikaner of Great Indian Desert. *TigerPaper*, 30(4): 3–8.
- Purohit, A.K., K.B. Vyas and K.R. Senacha (2006). Population dynamics of bats in and around Jodhpur of Great Indian Desert. *TigerPaper*, 33(3): 15–22.
- Purohit, A.K. and K.B. Vyas (2006). Review of sex ratio in several bat species inhabiting the Great Indian Desert. *Vespertilio*, 9-10: 233–235.
- Ramaswami, L.S. and T.C. Anand Kumar (1963). Differential implantation of twin blastocysts in *Megaderma* (Microchiroptera). *Cellular & Molecular Life Sciences*, 19(12): 641–642.
- Ryley, K. (1914). Bombay Natural History Society's Mammal Survey of India. No. 12. Palanpur and Mt. Abu. *Journal of the Bombay Natural History Society*, 22(4): 684–699.
- Senacha, K.R. and S. Dookia (2013). Geoffroy's Trident Leaf-nosed Bat, *Asellia tridens* (Geoffroy, E., 1813) from India. *Current Science*, 105(1): 21–22.
- Senacha, K.R. and A.K. Purohit (2004). Possible twin birth in the Indian flying fox *Pteropus giganteus*. *Bat Research News*, 45(4): 199.
- Senacha, K.R. and A.K. Purohit (2005). Partial albinism in the lesser mouse-tailed bat, *Rhinopoma hardwickii*. *Bat Research News*, 46(3): 75–76.
- Senacha, K.R., K.B. Vyas and A.K. Purohit (2006). New records of Short-nosed Fruit Bat *Cynopterus sphinx* (Vahl, 1797) from Thar Desert, Rajasthan. *Zoos'Print Journal*, 21(10): 2419–2420.
- Senacha, K.R. (2002). A note on *Prosopis julifora julifera*: Emerging threat for the microchiropteran of Thar Desert. *BatNet - CCINSA Newsletter*, 3(2): 7–8.
- Senacha, K.R. (2003). Eco-status and demographic changes among the chiropterans of the Thar desert with special reference to Jodhpur. Ph.D. Thesis, JNV University, Jodhpur, India. 175 pp.
- Senacha, K.R. (2006). Opportunistic survey of Indian Flying Fox *Pteropus giganteus* (Brunnich, 1782). *BatNet - CCINSA Newsletter*, 7(1-2): 27–29.
- Sharma, S.K. (1986). Painted bats and nests of Baya Weaver Bird. *Journal of the Bombay Natural History Society*, 81: 196.
- Simmons, N.B. and A.L. Cirranello, (2020). Bat Species of the World: A taxonomic and geographic database. Accessed on 08/06/2021.
- Sinha, Y.P. (1980). The bats of Rajasthan: taxonomy and zoogeography. *Records of the Zoological Survey of India*, 76(1-4): 7–63.
- Sinha, Y.P. (1981). New record of Black-bearded tomb bat, *Taphozous melanopogon melanopogon* Temminck from Rajasthan. *Geobios, Jodhpur*, 8(5): 225–226.
- Sinha, Y.P. (1996). Bats in Indian Thar Desert. pp. 349–352. *In*: Ghosh, A.K, Q.H. Baqri & I. Prakash (editors). *Faunal Diversity in the Thar Desert: Gaps in Research.* Scientific Publication, Jodhpur. p. i-xi, 1-140.

- Srinivasulu, C. and B. Srinivasulu (2006). Biodiversity Assessment Survey of Hadoti Region, Rajasthan. Part I. Birds and other wildlife of Jhalawar and Kota districts. Unpublished Report submitted to Maharana Pratap University of Agriculture and Technology, Udaipur. 24 pp.
- Srinivasulu, C., A. Srinivasulu and B. Srinivasulu (2021). Checklist of the bats of India (v1.2). https://threatenedtaxa.org/index.php/JoTT/checklists/bats/india [Date of publication: 19 May 2021].
- Srinivasulu, C., B. Srinivasulu and Y.P. Sinha (2013). Chiropteran fauna of Rajasthan: Taxonomy, distribution and status. pp. 505–548. *In*: Sharma, B.K., S. Kulshrestha & A.R. Rahmani (eds.). *Faunal Heritage of Rajasthan, India: General Background and Ecology of Vertebrates*. Springer Science+Business Media New York., USA.
- Topal, G. (1971). The taxonomic position of *Myotis dobsoni* (Trouessart, 1879) and some statistical data to the subspecific examination of *Myotis blythi* (Tomes, 857). *Annales Musei historico-naturalis Hungarici*, 63: 383–400.
- Wason, A. (1978). Observations on homing ability of some insectivorous bats. *Zeitschrift für Säugetierkunde*, 44(5): 305–306.
- Wilson, D. E. and R.A. Mittermeier (chief editors) (2019). *Handbook of the Mammals of the World. Vol. 9. Bats*. Lynx Edicions, Barcelona. 1008 pp. + 74 colour plates, 404 colour photographs, and 1422 distribution maps.
- Wroughton, R.C. (1918). Summary of the results from the Indian Mammal Survey of Bombay Natural History Society. I. Primates and Chiropetra (Part). *Journal of the Bombay Natural History Society*, 25: 547–598.

CHAPTER

6

An Eco-Friendly and Economical Method for Large Scale Manufacturing of Iron and Iron Oxide Nanoparticle Using Extract of Lac

Mr. Nitin Prakash Koche, Dr. Kantilal Hiralal Nagare and Dinesh Daujirao Wanule, Mr. Rinkesh Vishnu Kurkure

Indian Nationals

Department of Zoology,

Birla College of Arts, Science and Commerce,

Birla College Road, Kalyan-421304

ABSTRACT

In the present invention a novel method of large scale production of Iron and Iron Oxide nanoparticle is invented using extract of lac (an animal Secretary Product). Method is one easy, less ingredient involving, cheap, ecofriendly, less energy consuming and help in production of Iron and Iron Oxide nanoparticle on large scale. This is the first successful method of preparation of Iron and Iron Oxide nano particles from extract of lac and or waste product of lac industry.

The following specification particularly describes the invention and the manner in which it is

FIELD OF INVENTION

The field of invention relates to an eco-friendly and economical method for large scale manufacturing of Iron and Iron Oxide nanoparticle using extract of Lac.

Background of Invention

Meaning of Greek word Nano is dwarf and used as prefix for 10 '. In another word nanometre is billionth part of metre. Nanoparticle has a particle whose at least one dimension is less than 100 rim. Nanoparticles have unique physical characteristics and chemical properties and hence are used in hundreds of industries. Advancement in nanotechnology, make possible synthesis of nanoparticles of various shapes, sizes and kinds. Metallic nanoparticles have been used in various areas like Contrast agents for Magnetic Resonance Imaging (MRI) ,As drug carriers for target specific drug delivery, As gene carriers for gene therapy ,As therapeutic agents for hyperthermia based cancer treatments, As magnetic sensing probes for in-vitro diagnostics (IVD), As Nano adjuvant for vaccine and antibody production.

Iron oxide is very useful in MRI in which they enhance the quality of image. It can also be used to targeted molecular imaging probes in MRI

Manmade nanoparticles have been now synthesized on largescale for its industrial applications.

The current methods of synthesis of metal based nano particles are costly and time consuming.

Moreover, Lac is the scarlet resinous secretion of a number of species of lac insects, of which the most commonly cultivated species is *Kerria lacca.*

The extract of Lac is potent reducing agent which helps in reduction of metal ions to the respective nanoparticles, and such property of the Lac is being exploited in the present invention;

A brief description of the accompanying drawing,

Fig l. Ultra Violet Spectroscopic photograph of Iron and Iron Oxide nanoparticle obtained by the method of the present invention;

Fig. 2 is a Transmission Electron Microscopy (TEM) photograph of the Iron and Iron Oxide nanoparticle by the method of the present invention;

Fig. 3 is Energy Dispersive Spectroscopy (EDS) photograph of Iron and Iron Oxide nanoparticles obtained by the method of the present invention;

Fig. 4 FTIR Spectrum of Iron and Iron Oxide nanoparticles obtained by the method of the present invention;

Fig. 5; is a graph of XRD pattern of Iron and Iron Oxide nanoparticles obtain by the method of the present invention;

Objective of Invention

The object of present invention is to provide economic method for production of Iron and Iron Oxide Nano particles in large quantities.

Statement of Invention

Method of Eco-friendly and Economical Bulk Synthesis of Iron and Iron Oxide nanoparticle using extract of lac.

A Summary of Invention

A novel method of Synthesis of Iron and Iron Oxide nanoparticle is invented using extract of lac. Method is one easy, less ingredient involving, cheap, ecofriendly, less energy consuming and help in production of Iron and Iron Oxide nanoparticle on large scale. This is the first successful method of preparation of Iron and Iron Oxide Nano particles from the lac extract and or waste product of lac industry.

Detailed Description of the Invention

In a preferred embodiment of the invention a method for manufacturing Iron and Iron Oxide nanoparticle is provided by using the lac extract;

Lac is the scarlet resinous secretion of a number of species of lac insects, of which the most commonly cultivated species is *Kerria lacca.*

Cultivation begins when a farmer gets a stick (brood lac) that contains eggs ready to hatch and ties it to the tree to be infested. Thousands of lac insects colonize the branches of the host trees and secrete the resinous pigment. The coated branches of the host trees are cut and harvested as stick lac.

The harvested stick lac is crushed and sieved to remove impurities. The sieved material is then repeatedly washed to remove insect parts and other soluble material. The resulting product is known as seed lac. The prefix seed refers to its pellet shape. Seed lac which still contains 3—5% impurities is processed into shellac by heat treatment or solvent extraction.

According to present invention lac extract is obtalned from many plants serve as host plants for lac insect. *Butea monosperma* (Flame of the forest — Palas), *Acacia arabica* Willd (Babul), *Acacia catechu* Willd (Khair), *Cajanus cajan* (Linn.) Millsp.-(Arhar; Tur.), *Ficus benghalensis* Linn.-Bar (Banyan tree), *Ficus cunia* Buch.-Ham.-khunia; jharphali, Ficus lacor Buch.-Ham.-Pilkhan, *Ficus racemosa* Linn.-Gular, *Ficus religiosa* Linn. (Pipal), *Leea crispa* Linn.-Bnn-chalta, *Leea robusta* Roxb. — Galeni. *Schleichera oleosa* (Lour.) Oken. (Kusum), *Zizyphus mauritiana* Lam. (Ber) and *Zizyphus xylopyra* Willd. — Ghont *Samanea saman* rain tree or any other minor host plant or produced in or discarded by-product extract during production of shellac and any other associated products of Lac based industries.

The lac extract is prepared in a solvent includes water Milli-Q, polar & non-polar solvents selected from Pentane,cyclopentane,hexane,cyc1ohexane,benzene,tolune,1,4-Dioxane, chloroform, ether, ethyl acetate, acetone, Dimethyl foramide, Acetonitrile, Isopropanol, n- propanol, ethanol, methanol, chloroform, acetone, Acetic Acid ,Nitro methane or in any organic solvent and combination thereof.

An extract is prepared by adding 10 gm of lac in a solvent (water) or mixture of solvent and allow 24 hrs undisturbed for extraction and thereafter filtered. The filtrate is used in present invention for synthesis of Iron and Iron oxide nanoparticle.

In the present invention a method for synthesis of Iron and Iron oxide nanoparticle is by dissociating one or more Iron salts;

The Iron salt is selected from a group consisting FeC1 FeCl, $6H_2O$, $Fe(NO_3)_3$ $9H_2O$, Fe_2(50 4), in lac extract and allowing a mixture to react at a 25-40°C temperature.

The mixture of the Iron salts and solution turns from light yellow to brown colour initially and getting turbid then dark brown in nature.

The nanoparticles are obtained in the form of powder by heating in range of up to 400-900°C in furnace, for 40 min to l hrs.

According on aspect of the invention it is provided that a method economic and ecofriendly synthesis of Iron and Iron Oxide Nano particle;

The concentration of Iron salt is in the range of 0.1- l.0M;

In the economical aspect of the invention it is revealed that an expenditure towards eco- friendly synthesis of Iron and Iron Oxide Nano particles For 100 Gram is Rs.3,880/- in comparison to the FetO3 alpha, high purity 99.5+oZ, 30nm $159/100g - 10612.45 Rs. (http://www.us-nano.com/inc/sdetail/594) as on dated 11-10-2016 , Fe3O4, 98+0X, 20-30 nm which Need water dispersion with 20wt% $89/100g -5940 Rs (http://www.us- nano.com/inc/sdetail/234) as on dated l 1-10-2016

The price may subject to change with quantity of Iron and Iron Oxide nanoparticles production in bulk;

(b) The substrate used in synthesis method is natural and approved by U.S. Food and Drug Administration as nontoxic. A & B (2016) reported various methods like chemical / electrochemical precipitation, vapour deposition, atomic / molecular condensation, spray pyrolysis, laser pyrolysis, aerosol pyrolysis , chemical etching, thermal / lesser ablation has quite expensive and potentially hazardous to the environment in the sense of the using the hazardous solvents, non-deteriorating chemicals which remains in the environment causing deteriorating effect on environment; where as in contrast to the conventional method the present method uses the solvent such as water, alcohols.

Example;

1.0 Preparation of Lac extract;

Aqueous lac extract collected from host plant *Samanea saman* is prepared by treating 1.0 gm of lac into l00ml Milli-Q water and kept overnlght at ambient temperature/ room temperature ranging between 10°C to 40°C and filtering it by using filter paper, the filtrate obtained is used for the preparation of Ifon/iron oxlde nanoparticles from Iron salts;

2.0 Reduction/ Dissociation of Fe (NOT), $9H_2O$ in to Iron and Iron Oxide nanoparticles;

The filtrate obtained from the Lac extraction is added to the 1.0 M solution of Fe $(NO_3)_3$ $9H_2O$. This mixture is allowed to react at ambient temperature/ room temperature ranging between 10°C to 40°C for 24 hrs (optimized).

As reaction proceeds the mixture of the Iron salts and solution turns from light yellow to brown colour initially and getting turbid then dark brown in nature. The powder of Iron oxide nanoparticle was obtained by heating solution at 400-900°C for 40 to 60 min.

3.0 Reduction/ Dissociation of FeC1, $6H_2O$ in to Iron and Iron Oxide nanoparticles;

The filtrate obtained from the Lac extraction is added to the 1.0 M solution of FeCl, $6H_2O$. This mixture is allowed to react at ambient temperature/ room temperature ranging between 10°C to 40°C for 24 hrs (optimized).

As reaction proceeds the mixture of the Iron salts and solution turns from light yellow to brown colour initially and getting turbid then dark brown in nature. The powder of Iron oxide nanoparticle was obtained by heating solution at 400-900°C for 40 to 60 min.

4.0 Reduction/ Dissociation of Feb(SOA), in to Iron and Iron Oxide nanoparticles;

The filtrate obtained from the Lac extraction is added to the 1.0 M solution of $Fe_2(504)_3$ This mixture is allowed to react at ambient temperature/ room temperature ranging between 10°C to 40°C for 24 hrs (optimized).

As reaction proceeds the mixture of the Iron salts and solution turns from light yellow to brown colour initially and getting turbid then dark brown in nature. The powder of Iron oxide nanoparticle was obtained by heating solution at 400-900°C for 40 to 60 min.

5.0 Characterization of Iron and Iron Oxide nanoparticles;

The presence of Iron and Iron Oxide nanoparticle has been validated by analytical technique such as UV Visible Spectroscopy, Fourier Transmission Infrared spectroscopy (FTIR), Energy Dispersive Spectroscopy (EDS), Transmission Electron Microscopy (TEM). Size and Surface morphology was studied by using TEM made applying procedure.

Fourier Transmission Infrared Spectroscopy was subjected to IR Source of range 4000 to 400 cm 1. Jasco FT/IR-4100 Type A model. Biosynthesis of Iron/Iron oxide nanoparticle was studied using UV Visible Spectroscopy SICAN-2301 of INCARP model. The absorbance peaks of nanopartlcle were found at 216 rim and 277 nm in UV range where 2l6nm is observed for Fe and 277 rim for Fe" the more sharp peak shows more no. of Fe atoms and are homogeneous in nature whereas the peak at 277 nm is broad and shows less homogeneous no. of nanoparticles, where the broadening of the peak is observed due to oxygen bonding ,as the no. of oxygen atoms reacting with Fe the probable states observed are FeO, Fe 03 and Fei O4: where Fe3O4 is generated by combination of FeO and Fe2O3 with respect to reaction temperature and wt% of oxygen. The energy band gap for Iron and Iron Oxide nanoparticles was experimentally calculated by using Planck—Einstein relation to be 5.74 eV.

TEM of Example l of Iron and Iron Oxide showed in fig. 2 Heavy globular agglomeration of nanoparticles of size 30 nanometres to 60 nanometre.

EDS of Example 1 of Iron and Iron Oxide shown in fig. 3 ensures formation of Iron and Iron Oxide nanoparticle.

FTIR analysis of Iron and Iron Oxide shown in fig. 4 shows prominent peak of Iron and Iron Oxide nanoparticle at 548.649 cm '. Stretchlng vibration of Iron and Iron Oxide Bond.

Claims

1. An economical, eco-friendly method for Large scale manufacturing of Iron oxide nanoparticles by using Lac extract comprising
   (a) providing the extract of lac in solvent at the ambient temperature;
   (b) reacting the extract of lac with standard Iron salt solutlon at kept overnight at ambient temperature/ room temperature ranging between l0°C to 40°C for 24 hrs;
   (c) roasting the Iron and Iron Oxide nanoparticle in step (b) heating solution up to 400-900°C for 60-40 min;
2. An economical, eco-friendly method for manufacturing of Iron and Iron Oxide nanoparticles by using Lac extract according to claim 1 wherein lac extract is prepared from lac produced by lac insect occurring on host plants.
3. An economical, eco-friendly method for synthesis of Iron and Iron Oxide nanoparticles by using Lac extract according to claim l or 2 plant host includes *Butea monosperma* (Flame of the forest — Palas), *Acacia arabica* Willd (Babul), *Acacia catechu* Willd (Khair), *Cajanus cajan* (Linn.) Mi1lsp.-(Arhar ; Tur.), *Ficus benghalensis* Linn.-Bar (Banyan tree), *Ficus cunia* Buch.-Ham.-khunia ; jharphali , *Ficus lacor* Buch.-Ham.-Pilkhan, *Ficus racemosa* Linn.-Gular, *Ficus religiosa* Linn.(Pipa1), *Leea crispa* Linn.-Bnn-chalta, *Leea robusta* Roxb. — Galeni. *Schleichera oleosa* (Lour.) Oken. (Kusum), *Zizyphus mauritiana* Lam. (Ber) and *Zizyphus xylopyra* Willd. —Ghont *Samanea saman* or any other source of lac.
4. An economical, ecofriendly method for synthesis of Iron and Iron Oxide nanoparticles by uslng Lac extract according to claim 1 wheieln solvent is selected from Milli-Q water ,any polar & nonpolar solvents, ether, ethanol, methanol, chloroform, acetone , or in any organic solvent and combination / mixture there of;
5. An economical, eco-friendly method for synthesis of Iron and Iron Oxide nanoparticles by using Lac extract according to claim 1 wherein Iron salt is selected from FeCl,, FeCl, $6H_2O$, Fe $(NO_3)_3$ $9H_2O$, $Fe_2(SO_4)_3$

6. An economical, eco-friendly method for synthesis of Iron and Iron Oxide nanoparticles by using Lac extract according to claim 1 whereln standard of Iron salt is 0.1 pM upto 1.0M or any other concentration;

Signature of applicants

Dr. Dinesh D.Wanule, Mr. Rinkesh V.Kurkure, antilal H. Nagare Mr. Nitin P. Koche

Figure and Drawing

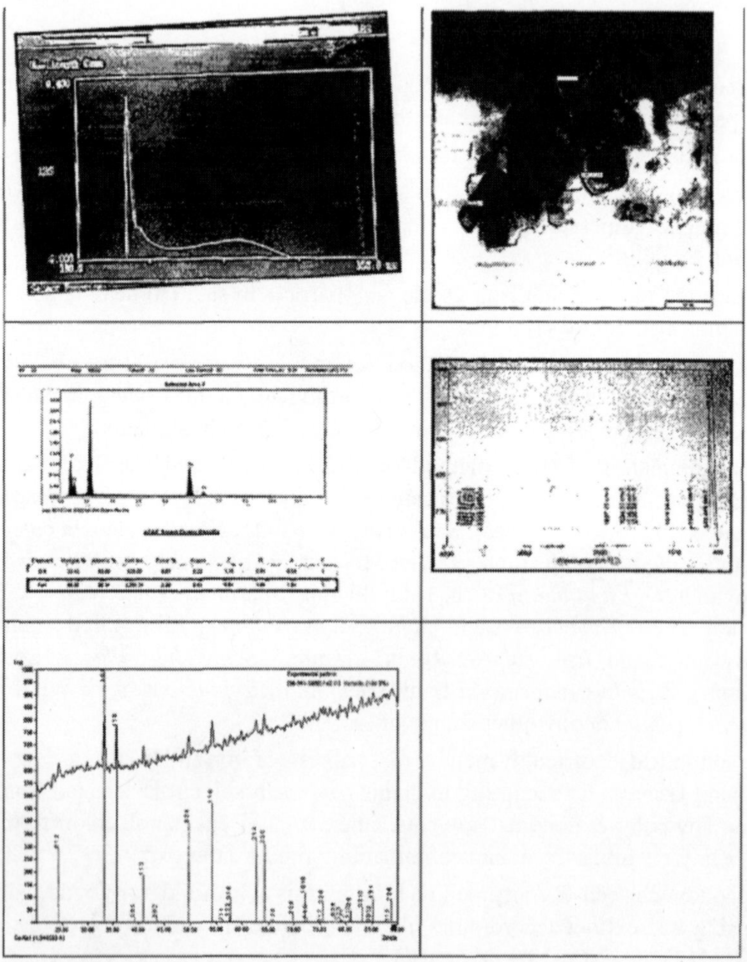

CHAPTER

7

An Update on Therapeutic Approach of Diabetes Mellitus through DPP-4 Inhibition

Karishma Sen and Heera Ram
Department of Zoology, Jai Narain Vyas University, Jodhpur (Rajasthan)-342011, India
E-mail-baradhr@gmail.com

ABSTRACT

The DPP-4 inhibition-based therapeutics of diabetes is one of the leading approach which governed by internal stimulation to pancreas through promoting the GIP (Glucose-dependent insulinotropic Peptide) and GLP-1(Glucagon like Peptide-1). The DPP-4 is a serine protease which promotes the degradation of GLP-1. The NCBI (National Centre for Biotechnology Information) search engine of protein filtered with gene DPP-4 says Gene ID: 1803 having RefSeq transcripts(8) RefSeq proteins(4) and PubMed(521) with different search of Orthologs, Genome Data Viewer, BLAST and the download option of the file with 1756 search options. The DPP-4 inhibitors or inhibitors like compounds are promoting the longevity which stimulates the pancreas by following the different mechanisms. The DPP-4 inhibitors are known as the gliptins which introduced in 2006 and nowadays more than 20 kinds of gliptins available in market as well as under the clinical trials. The DPP-4 inhibition governed by following the uses of clinical gliptins and applications of nutraceuticals. Beside this, the gliptins shown to neutral body weight, demonstrated cardiovascular safety and other pleotropic activities. Along with this, the nutraceutics is enhancing the incretin endocrinal activities following through the analogues of GLP-1, DPP-4 inhibitors as well as free radical scavenging activities. Therefore, the focus is to make provision of the DPP4 inhibitors with minimal side effects for the treatment of type 2 diabetes. This chapter is summarising the DPP-4 inhibition-based strategies for therapeutics of diabetes.

INTRODUCTION

DPP-4 was first characterized by Hopsu-Havu and Glenner. DPP-4 is type 2 membrane peptidase resembling cluster of differentiation 26(CD-26), marker of activated T-lymphocytes(Deacon, 2019; Drucker, 2013)also known as the T-cell antigen CD26, is a multi-functional protein which, besides its catalytic activity, also functions as a binding protein and a ligand for a variety of extracellular molecules. It is an integral membrane protein expressed on cells throughout the body, but is also shed from the membrane and circulates as a soluble protein in the plasma. A large number of bioactive molecules can be cleaved by DPP-4 in vitro, but only a few of these have been demonstrated to be physiological substrates. One of these is the incretin hormone, glucagon-like peptide-1 (GLP-1. DPP-4 is ubiquitously expressed glycoprotein of 110 kDa. It is the member of proteases family also known as Adenosine deaminase (ADA) binding protein. It is the membrane associated peptidase which comprises of 26 exons that encode 766 amino acids.

DPP-4 gene is located on long arm of chromosome 2 (2q24.3) (Li *et al.*, 2018; Scheen, 2015). It is an ectopeptidase involved in the degradation of several peptides and hormones including glucagon family peptides, chemokines, and neuropeptides from the N-terminal of polypeptide. It is present in numerous tissues, T-cells, B-cells and natural killer cells, carries important immune functions. It is widely distributed in the organs such as liver, gut, endothelial capillaries, acinar cells of mucous and salivary glands, pancreas, uterus with the highest level found in kidney proximal tubules, brush border of the small intestine, bone marrow and immune organs such as thymus, spleen and lymph node(Kawanami *et al.*, 2021). It also exists as a soluble circulating form in plasma and tethered to endothelium and epithelial cells especially in the intestinal mucosa. DPP-4 is an intrinsic membrane glycoprotein and a serine exopeptidase which cleaves from the penultimate position of X-proline or X-alanine dipeptides from the N-terminal or amino terminus of polypeptides(Andersen *et al.*, 2018; Green *et al.*, 2007).Both catalytic and non-catalytic functions are performed by DPP-4. Catalytic functions involve as N-terminal dipeptidase and leads to the degradation of incretin hormones and other peptides whereas non catalytic functions involve in the interaction of peptide with multiple ligands including T-cell co stimulation, ADA and fibronectin binding functions in immune regulation(Shawky *et al.*, 2020)a dipeptidyl peptidase-4 (DPP-4. DPP-4 exists as a dimmer both *in vivo* as well as*in vitro*as shown by the studies of cross-linking and ultracentrifugation. Homodimerization of DPP-4 is necessary for the activity of serine protease. DPP-4 is the canonical representative of DPP-4 gene family (S9 family) of genetically related peptidase which includes DPP-8, DPP-9, FAP (Fibroblast activation protein), DPP-6 and DPP-10. DPP-6 and DPP-10 are catalytically inactive protein. DPP-4 gene family shows the structural similarity due to which DPP-4 could be the inhibitors of other enzymes leading to several side effects. Physiological substrates of DPP-4 are extensive and contain alanine or proline containing peptides which includes substance P, neuropeptide Y, peptide YY, and growth hormone-releasing hormone (GHRH). Over the last few years, Inhibition of DPP-4 is one of the approaches and has vast importance for the treatment of T2DM as it enhances the glucose dependent insulin secretion. DPP-4 inhibitors have revealed high selectivity profiles and pharmacokinetic properties(Deacon, 2019; McCloskey *et al.*, 2020).DPP-4isatransmembraneglycoproteinthatformsahomodimer or tetramer on the plasma membrane.DPP-4 transmits signals across cell membranes and interacts with other membrane proteins.In terms of amino acid sequence, DPP-4 activity ishighlyconserved among species. DPP-4 molecule consists of four domains of N-terminal cytoplasmic domain (AA1–6) of 6 amino acid extending into cytoplasm, 22-residue transmembrane domain (TMD) (AA7–28), a flexible stalk segment (AA29-39) andlargeC-terminalextracellulardomain (AA40-766). Extracellular protein includes the C-terminal catalytic domain, a cysteine-rich area, and a large, glycosylated region linked by a flexible stalk to the 22-amino acid transmembrane domain. Extracellular domain involves in the dipeptidyl peptidase activity and binding to the ligands such as Adenosine deaminase (ADA) and fibronectin.

Residue AA294 and residues AA340–343 within thecysteine-rich segment are essential for ADA binding.The catalytic domain (AA508-766) consists of the amino acid triad Ser630, Asp708 and His740.DPP-4 molecule consists of two binding pockets called as S1 pocket and S2 pocket. The S1 pocket consists of hydrophobic substituent having the catalytic triadSer630, Asp708 and His740 as side chain and has specificity of proline amino acid. The S2 pocket consists of aromatic or heterocyclic rings having residues Arg125,

Glu205, Glu206, Phe357, Ser209 and Arg358(Enz *et al.*, 2019; Xing *et al.*, 2014). The core scaffold is the active site having primary amine, an aromatic ring and variable constituent. S2 pocket has inhibitor selectivity of DPP-4 over other DPP-4 gene family members. S2 pocket residues Glu205 and Glu206 have shown involvement in hydrogen bondingor salt bridge interactions with the cationic groups. DPP-4 inhibitor occupies similar S1-S2 space but does not interact with catalytic serine and does not affect the structure of DPP-4(Deacon, 2019).

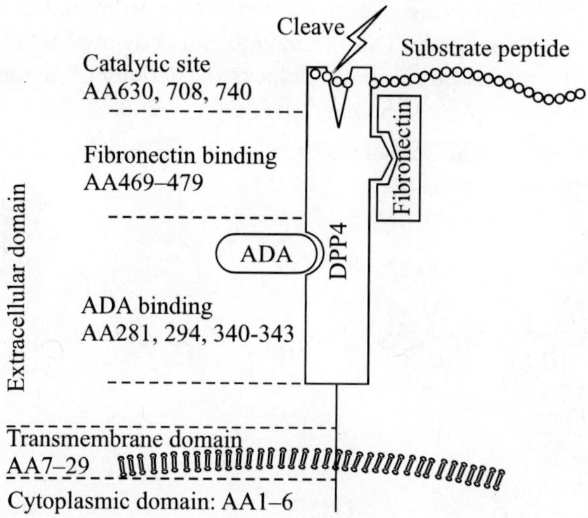

Fig. 1. *Molecular Structure of DPP-4(Pantaleão et al., 2018)).*

Incretins

The word "incretine" was coined by the Belgian physiologist Jean La Barre in 1932. Incretin is the gut hormone released from gastrointestinal tract in response to nutrient ingestion mainly fat and glucose that further stimulates the secretion of pancreatic hormones(Meier & Nauck, 2015).Mentlein *et al.* in 1993 found two incretin hormones as substrate of DPP-4 enzyme which includes GLP-1(Glucagon like peptide 1) and GIP (Glucose dependent insulinotropic polypeptide)(Rehfeld, 2018; Salvatore *et al.*, 2019). Bothhave significant effects on insulin secretion, glucose regulation and share considerable sequence homology. GLP-1 and GIP both are the members of the glucagon peptide super family and maintain homeostasis. Competitive inhibition of DPP-4enhances the bioavailability as well as half life of active incretin hormones which further leads to increase in their biological effect(Girard, 2008; Yabe & Seino, 2013).

GIP (Glucose Dependent Insulinotropic Polypeptide or Gastric Inhibitory Peptide)

Inearly1970's, GIP was the first incretin hormone isolated from the crude extracts of porcine small intestine inthelaboratory of Viktor Mutt in Stockholm, John Brown as an inhibitor of gastric acid secretion and named as gastric inhibitory peptide. It was renamed as Glucose dependent insulinotrophic polypeptide due to its role and physiological action as incretin hormone and has glucose stimulated insulin secretion(Meier & Nauck, 2015; Mudaliar & Henry, 2012). It is secreted by neuro-endocrine K-cells present in stomach and proximal small intestine (duodenum and jejunum). It is a single 42 amino acid peptide with molecular weight 4984 g/mol derived from a larger 153 amino acid precursor (Figure 2). Meal rich in carbohydrate and lipid increases the circulating plasma levels of GIP by 10-20 folds. It enhances the secretion of insulin only in the presence of high concentration of glucose. Amino acid sequence of GIP is highly conserved across species with over 90% homology. It does not inhibit glucagon secretion or gastric emptying. Half life of intact GIP is approximately 7 minutes in non-diabetic person and 5 minutes in diabetic patients(Drucker, 2006). Following secretion into the circulation, intact GIP (1-42 amide) is cleaved from the N- terminus by DPP-4 enzyme and converted into inactive truncated GIP (3-42 amide). The inactive truncated GIP lacks the activity of incretin and may act as competitor of GIP at its receptor(Al-Sabah, 2016; Girard, 2008).

Sequence of GIP: YAEGXFXSDYSXAMDKXHQQDFVNWLLAQKGKKNDWKHNXXQ (Obtained from PubChem)

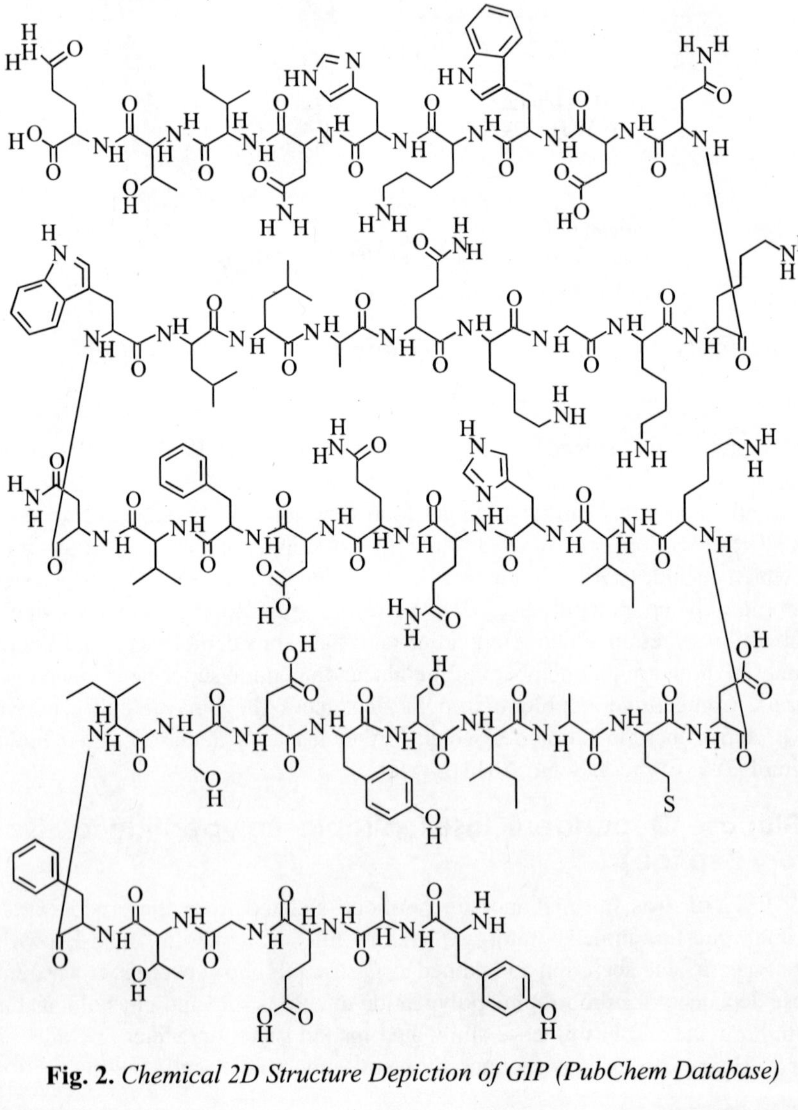

Fig. 2. *Chemical 2D Structure Depiction of GIP (PubChem Database)*

GLP-1(Glucagon Like Peptide-1)

In early 1980s, GLP-1a product of glucagon gene was first identified(Holst, 2019). It is generally secreted by neuro-endocrine L-cells present in the distal small intestine (ileum and colon) and large intestine. Both neural and endocrine cells control the secretion of GLP-1(Di Dalmazi *et al.*, 2012). It circulates in two equipotent molecular forms: GLP-1 (7-37) amide and GLP-1(7-36) amide and collectively referred to as intact bioactive GLP-1. GLP-1 (7-36) amide constitutes 80% of circulating GLP-1(MacDonald *et al.*, 2002; Yabe & Seino, 2013). Since, it is cleared by the kidney and rapidly degraded by the ubiquitous enzyme DPP-4, half life is less than 2 minutes.As a result of DPP-4 activity, only 10–20% of total plasma GLP-1 is biologically active. The primary physiological responses include inhibition of glucagon secretion, enhance glucose-dependent insulin secretion, induce satiety, inhibition of gastric acid secretion, and increase islet cell mass leads to lowering in blood glucose levels.It also influences the regulation of appetite; delays gastric emptying and can result in weight loss which is highly beneficial in the treatment of T2DM(Villalba *et al.*, 2020). GLP-1 with molecular weight 580g/mol has antidiabetic action; hence preventing its inactivation by DPP-4 enzyme and further used as a novel approach for the treatment ofT2DM.Both GLP-1 and GIP have structurally distinct yet overlapping actions and exert their actions through specific G-protein coupled receptors. They both are degraded by DPP-4 enzyme converting them from active to inactive metabolites. GLP-1 is degraded even before leaving the gut as DPP-4 molecules are anchored to the luminal surface of the endothelial cells of the mucosal capillaries(Meier & Nauck, 2015; Waget *et al.*, 2011). GIP leaves the gut un-degraded due to less susceptibility towards DPP-4. Other peptides and chemokines are also cleaved by DPP-4 activity. GLP-1 sequence is not available on the PubChem Database.

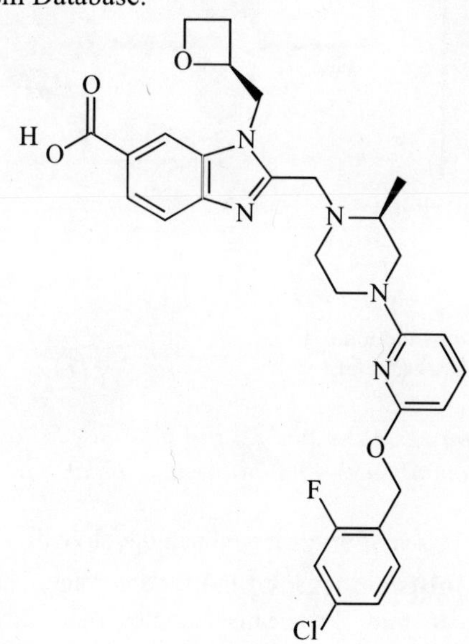

Fig. 3. *Chemical 2D Structure Depiction of GLP-1 (PubChem Database)*

Mechanism of Action

In response to the ingestion of food, the glucose level in blood increases due to the presence of enzymes in the gastrointestinal tract, which digests food(Gupta & Kalra, 2011; Villalba *et al.*, 2020; Zhao *et al.*, 2017). In response to glucose, incretin hormones such as GLP-1 and GIP are secreted from the small intestine, further rapidly degraded by the enzyme DPP-4. DPP-4 enzyme is responsible for metabolising incretin hormones. DPP-4 inhibitor inhibits DPP-4 enzyme which prevents the degradation of GLP-1 and GIP and enhances glucose-stimulated insulin secretion from the pancreas (incretin action). DPP-4 inhibitors activity results in meal-based enhancement of GLP-1 and GIP.GLP-1 and GIP act on the pancreatic β-cell to increase post-prandial insulin release in a glucose dependent manner(Nicolucci& Rossi, 2008). GLP-1 also acts on the pancreatic α-cell to suppress glucagon release leading to decreased hepatic glucose production. The increased cellular glucose uptake and the decreased hepatic glucose output helps in maintaining the normal glucose level. In type 2 diabetes patients, meal stimulated circulating levels of GLP-1 is reduced and GIP secretion is impaired(Al-Sabah, 2016; Alonso *et al.*, 2012). The utility of GLP-1 as a pharmacologic agent is limited due to its short plasma half life. Therefore, DPP-4 inhibitors are one of the latest pharmaceutical targets for type 2 diabetes treatment.

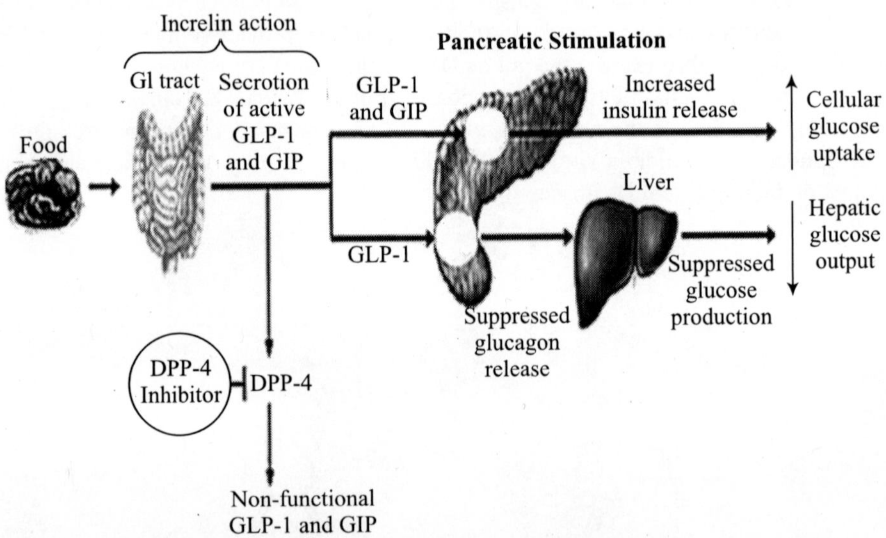

Fig. 4. *Mode of action of DPP-4 and DPP-4 inhibitor (Adopted from Diabetes in control -Role of DPP-4 inhibitors in clinical practice article)*

Based on GLP-1 two classes of drugs have been developed:

1. **GLP-1 receptor agonists/analogs:** GLP-1 receptor agonists have given importance by both the American and European Diabetes Associations from 2012, which recommend the use of incretin mimetics(Chia & Egan, 2008; Waget *et al.*, 2011). They bind to the GLP-1 receptor and mimetics the endogenous action of GLP-1. They

are peptide-based therapies, injectable, more potent glucose lowering agents having single G protein coupled receptor target.They have shown narrow pharmacokinetic profile, but they can achieve higher levels of GLP-1. Drugs have shown tolerable side effect profile likenausea and vomiting. Overdose of drugs can become problematic. They cause weight reduction and rarely cause hypoglycaemia as they enhance insulin production only when there is increase in the levels of glucose(McCloskey *et al.*, 2020). Drugs include in GLP-1 receptor agonists are Liraglutide, Exenatide, Lixisinatide. Two drugs of GLP-1 receptor agonists are licenced by the U.S., Europe and Japan includes exenatide.

2. **DPP-4 Inhibitors:** DPP-4 inhibitors are small molecular weight drugs approved by the U. S. Food and Drug Administration (FDA). They have "gliptin" as a suffix in their name and are incretin enhancers(Gallwitz, 2019; Sneha & Doss, 2016). Gliptins are the new class of antidiabetic drug that inhibit the proteolytic activity of DPP-4 and enhances the incretin effect. They are orally administered, potent and selective DPP-4 inhibitor having multiple G-protein-coupled targets. The gliptins enhance the levels of the incretin hormones, GLP-1 and GIP and reduces the plasma activity of DPP-4 enzyme by 70-90%. They have improved glycaemic controls which have an increasing role in the management of T2DM. DPP-4 inhibitors can be used in the treatment of T2DM either as monotherapy or as combinational therapy with other oral antihyperglycaemic agents like metformin, thiazolidinediones (TZDs) except GLP-1 analogs. Gliptins produce similar reductions in blood glucose glycated haemoglobin (HbA1c) levels, but they serve as several attractive clinical advantages. DPP-4 inhibitors are designed on the basis of tolerability, potency, efficacy and safety with minimal side effects. They are competitive reversible inhibitors of DPP-4 substrate. The first approved DPP4 inhibitor is sitagliptin in 2006. Other DPP-4 inhibitors such as vildagliptin, saxagliptin, alogliptin, linagliptin, anagliptin and teneligliptin were introduced latter(Kushwaha et al, 2014; Scheen, 2015). Vildagliptin, saxagliptin, sitagliptin, and alogliptin are approved to be antidiabetic agents by the United States and Europe. These drugs are well tolerated, weight neutral and have low risk of hypoglycaemia. DPP-4 inhibitors reduced HbA1c by 0.7% when used as a monotherapy or in combination with other agents.

Currently Available Some Approved DPP-4 Inhibitors

1. Sitagliptin (developed by Merck Sharp and Dohme Corp, approved as Januvia by US FDA in year 2006)
2. Vildagliptin, LAF-237 (developed by Novartis, approved as Galvus by EU in year 2007)
3. Saxagliptin, BMS-477118 (developed by Bristol-Myers Squibb, approved as Onglyza by US FDA in 2010)
4. Linagliptin, BI-1356 (developed by Boerhinger Ingelheim, approved as Tradjenta by US FDA in year 2011)
5. Alogliptin, SYR-322 (developed by Takeda Pharmaceutical Company Limited, approved for use in Japan)

6. Gemigliptin, LC15-0444 (developed by LG Life sciences Ltd, Korea approved by Korean FDA in June 2012)
7. Teneligliptin, MP-0513 (developed by Mitsubishi Tanabe and approved for use in Japan)
8. Anagliptin, SK-0403 (developed by Sanwa Kagaku Kenkyusho and approved for use in Japan)

DPP-4 Inhibitors Undergoing Clinical Trails

1. BI 1356 BS (undergoing phase 2 clinical trial and being developed by Boehringer Ingelheim)
2. Melogliptin (undergoing phase 2 clinical trial and being developed by Glenmark Pharmaceuticals Limited)
3. AMG 222 (undergoing phase 2 clinical trial and being developed by Amgen)
4. MP 513 (undergoing phase 2 clinical trial and being developed by Mitsubishi Tanabe Pharma Corporation)
5. Dutagliptin, PHX 1149 (undergoing phase 2 clinical trial and being developed by Phenomix Corporation)
6. PSN 9301 (undergoing phase 2 clinical trial and being developed by Prosidion)
7. Carmegliptin,R 1579 (undergoing phase 2 clinical trial and being developed by F. HoffmannLa Roche)
8. Trelagliptin, SYR 472 (undergoing phase 2 clinical trial and being developed by Takeda San Diego)
9. TA 6666 (undergoing phase 2 clinical trial and being developed by Mitsubishi Tanabe Pharma Corporation)
10. Denagliptin, GSK-823093 (undergoing phase 3 clinical trial and being developed by Glaxo Smithkline)
11. Gemigliptin (undergoing phase 3 clinical trial and being developed by LG Life Sciences)
12. Omarigliptin, MK-3102 (undergoing phase 3 clinical trial and being developed by Merck)
13. Gosogliptin, PF-00734200 (being developed by Pfizer)
14. Retagliptin, SP-2086 (undergoing phase 3 clinical trial and being developed by Jiangsu Hengrui Medicine)
15. Evogliptin, DA-1229(completed phase 2 clinical trial and being developed by Dong-A Pharmaceutical in Korea)

Side Effects of DPP-4 Inhibitors

The aim in pharmaceutical research is either to inhibit DPP-4 or to use the compounds that can partially resist DPP-4 activity. DPP-4 is not specific for GLP-1 and it has affinity for wide range of substrates. It has the potential to produce multiple effects from a single gene

Selective inhibition of DPP-4 enzyme should be done as it also shows catalytic activity with other peptides. DPP-4 activity leads to biologically alterations in the endogenous levels of these physiological substrates which lead to several adverse side effects. Mostly synthetic DPP4 inhibitors are well-tolerated but prolonged usage can cause adverse effects such as pancreatitis, angioedema, infective disorders, pancreatic cancer, and thyroid cancer, nasopharyngitis, headaches and urinary infections (Lamos et al., 2019; Tseng et al., 2015). Several other side effects such as severe hypoglycemia, lactic acidosis, idiosyncratic livercell injury, permanent neurological deficit, digestive discomfort, dizziness and even death(Lamos et al., 2019; Vangoitsenhoven et al., 2012). Continuous exposure of insulin leads to decreasein the number of receptors on the cell surfaceand causes degradation of hormone occupied receptors.

CONCLUSION

DPP4 inhibitors are a new class of pharmaceutical agents having improved glycaemic control and are well tolerated with neutral weight gain and low chances of hypoglycaemia. DPP4 inhibition prolongs the endogenous action of incretins GLP-1 and GIP which further enhances the glucose-stimulated insulin secretion and maintains the glucose homeostasis. Drugs used in the treatment of T2DM have several adverse side effects and the prohibitive cost is also one of the limiting factors associated with this disease. By using non-selective DPP4 inhibitors it was found that DPP4 plays an important role in the control of immune regulation, transplantation biology, cancer cell growth and metastasis. Therefore, highly selective DPP4 inhibitors with minimal side effects, low cost must be generated and developed for the treatment of type 2 diabetes. In modern days, herbal based phytomedicines are used worldwide to treat diabetes. The demand of phytomedicines increases due to its availability, low cost and fewer side effects. The scientific validations of medicinal plants are rapidly developing with distinctive pharmacological mechanism of action which are further used in the treatment and management of diabetes.

REFERENCES

- Al-Sabah, S. (2016). Molecular Pharmacology of the Incretin Receptors. *Medical Principles and Practice*, *25*(1), 15–21.
- Alonso, N., Teresa Julián, M., Puig-Domingo, M., & Vives-Pi, M. (2012). Incretin hormones as immunomodulators of atherosclerosis. *Frontiers in Endocrinology*, *3*(SEP), 1–6.
- Andersen, E. S., Deacon, C. F., and Holst, J. J. (2018). Do we know the true mechanism of action of the DPP-4 inhibitors? In *Diabetes, Obesity and Metabolism* (Vol. 20, Issue 1).
- Chia, C. W., and Egan, J. M. (2008). Incretin-Based Therapies in Type 2 Diabetes Mellitus. *The Journal of Clinical Endocrinology & Metabolism*, *93*(10), 3703–3716.
- Cho, N. H., Shaw, J. E., Karuranga, S., Huang, Y., da Rocha Fernandes, J. D., Ohlrogge, A. W., & Malanda, B. (2018). IDF Diabetes Atlas: Global estimates of diabetes prevalence for 2017 and projections for 2045. *Diabetes Research and Clinical Practice*, *138*, 271–281.

- Cho, Y. M. (2017). Characteristics of the pathophysiology of type 2 diabetes in Asians. *Annals of Laparoscopic and Endoscopic Surgery*, *2*, 14–14.
- Claire M. Steppan, Shannon T. Bailey, Savitha Bhat, Elizabeth J. Brown, Ronadip R. Banerjee, Christopher M. Wright, Hiralben R. Patel, R. S. A. and M. A. L. (2001). The hormone resistin links obesity to diabetes. *Nature*, *409*(6818), 307–312.
- Daiji Kawanami, Yuichi Takashi, Hiroyuki Takahashi, R. M. and M. T. (2021). Renoprotective Effects of DPP-4 Inhibitors. *Antioxidants*, *10*(246), 1–17.
- Deacon, C. F. (2019). Physiology and Pharmacology of DPP-4 in Glucose Homeostasis and the Treatment of Type 2 Diabetes. *Frontiers in Endocrinology*, *10*, 1–14.
- Di Dalmazi, G., Pagotto, U., Pasquali, R., and Vicennati, V. (2012). Glucocorticoids and Type 2 Diabetes: From Physiology to Pathology. *Journal of Nutrition and Metabolism*, *2012*, 1–9.
- Drucker, D. J. (2006). The biology of incretin hormones. *Cell Metabolism*, *3*(3), 153–165.
- Drucker, D. J. (2013). Incretin action in the pancreas: Potential promise, possible perils, and pathological pitfalls. *Diabetes*, *62*(10), 3316–3323.
- Enz, N., Vliegen, G., De Meester, I., & Jungraithmayr, W. (2019). CD26/DPP4 - a potential biomarker and target for cancer therapy. *Pharmacology and Therapeutics*, *198*, 135–159.
- Fonseca, V. A. (2009). Defining and characterizing the progression of type 2 diabetes. *Diabetes Care*, *32 Suppl 2*, S151–S156.
- Gallwitz, B. (2019). Clinical use of DPP-4 inhibitors. *Frontiers in Endocrinology*, *10*(JUN), 1–10.
- Girard, J. (2008). The incretins: From the concept to their use in the treatment of type 2 diabetes. Part A: Incretins: Concept and physiological functions. *Diabetes and Metabolism*, *34*(6), 550–559.
- Green, B. D., Flatt, P. R., & Bailey, C. J. (2007). Dipeptidyl peptidase IV (DPP IV) inhibitors: a newly emerging drug class for the treatment of type 2 diabetes. *Diabetes and Vascular Disease Research*. 3(3)159 – 165.
- Gupta, V., & Kalra, S. (2011). Choosing a Gliptin. *Indian Journal of Endocrinology and Metabolism*, *15*(4), 298–308.
- Heller, S., & Novodvorsky, P. (2019). Hypoglycaemia in diabetes. In *Medicine (United Kingdom)* (Vol. 47, Issue 1). 47(1), 1 – 10.
- Holst, J. J. (2019). From the incretin concept and the discovery of GLP-1 to today's diabetes therapy. *Frontiers in Endocrinology*, *10*(APR), 1–10.
- Kaneto, H. (2015). Pathophysiology of type 2 diabetes mellitus. *Nihon Rinsho. Japanese Journal of Clinical Medicine*, *73*(12), 2003–2007.

- Lamos, E. M., Hedrington, M., & Davis, S. N. (2019). An update on the safety and efficacy of oral antidiabetic drugs: DPP-4 inhibitors and SGLT-2 inhibitors. *Expert Opinion on Drug Safety*, *18*(8), 691–701.
- Lankatillake, C., Huynh, T., & Dias, D. A. (2019). Understanding glycaemic control and current approaches for screening antidiabetic natural products from evidence-based medicinal plants. *Plant Methods*, *15*(1), 1–35.
- Li, N., Wang, L. J., Jiang, B., Li, X. qian, Guo, C. long, Guo, S. ju, & Shi, D. Y. (2018). Recent progress of the development of dipeptidyl peptidase-4 inhibitors for the treatment of type 2 diabetes mellitus. *European Journal of Medicinal Chemistry*, *151*, 145–157.
- MacDonald, P. E., El-kholy, W., Riedel, M. J., Salapatek, A. M. F., Light, P. E., & Wheeler, M. B. (2002). The multiple actions of GLP-1 on the process of glucose-stimulated insulin secretion. *Diabetes*, *51*(SUPPL. 3), S434–S442.
- McCloskey, A. G., Miskelly, M. G., Flatt, P. R., & McKillop, A. M. (2020). Pharmacological potential of novel agonists for FFAR4 on islet and enteroendocrine cell function and glucose homeostasis. *European Journal of Pharmaceutical Sciences*, *142*, 105104.
- Meier, J. J., & Nauck, M. A. (2015). Incretin-based therapies: where will we be 50 years from now? *Diabetologia*, *58*(8), 1745–1750.
- Mudaliar, S., & Henry, R. R. (2012). The incretin hormones: From scientific discovery to practical therapeutics. *Diabetologia*, *55*(7), 1865–1868.
- Nicolucci, A., & Rossi, M. C. (2008). Incretin-based therapies: a new potential treatment approach to overcome clinical inertia in type 2 diabetes. *Acta Bio-Medica : Atenei Parmensis*, *79*(3), 184–191.
- Mudgil, P., Kamal, H. and Yuen, G.C. and Maqsood, S. (2018).
- Mudgil L.P., Kamal, H. and Yuen, G.C. and Maqsood, S. (2018). Characterization and identification of novel antidiabetic and anti-obesity peptides from camel milk protein hydrolysates. *Food Chemistry*. https://doi.org/10.1016/j.foodchem.2018.03.082
- Pantaleão, S. Q., Philot, E. A., de Resende-Lara, P. T., Lima, A. N., Perahia, D., Miteva, M. A., Scott, A. L., & Honorio, K. M. (2018). Structural dynamics of DPP-4 and its influence on the projection of bioactive ligands. *Molecules*, *23*(2), 1–10.
- Ram Najar Kushwaha, W. H. and S. B. K. (2014). Discovery of 17 Gliptins in 17-Years of Research for the Treatment of Type 2 Diabetes: A Synthetic Overview. *Chemistry & Biology Interference*, *4*(3), 137–162.
- Rehfeld, J. F. (2018). The origin and understanding of the incretin concept. *Frontiers in Endocrinology*, *9*(JUL), 1–7.
- Röder, P. V., Wu, B., Liu, Y., & Han, W. (2016). Pancreatic regulation of glucose homeostasis. *Experimental & Molecular Medicine*, *48*(November 2015), e219 (1-19).

- Salvatore, T., Nevola, R., Pafundi, P. C., Monaco, L., Ricozzi, C., Imbriani, S., Rinaldi, L., & Sasso, F. C. (2019). Incretin hormones: The link between glycemic index and cardiometabolic diseases. *Nutrients*, *11*(8), 1–12.
- Scheen, A. J. (2015). A review of gliptins for 2014. *Expert Opinion on Pharmacotherapy*, *16*(1). 1 – 10.
- Shawky, L. M., Morsi, A. A., Bana, E. El, & Hanafy, S. M. (2020). The biological impacts of sitagliptin on the pancreas of a rat model of type 2 diabetes mellitus: Drug interactions with metformin. *Biology*, *9*(1), 1–27.
- Sneha, P., & Doss, C. G. P. (2016). Gliptins in managing diabetes - Reviewing computational strategy. *Life Sciences*, *166*, 108–120.
- Tseng, C., Lee, K., & Tseng, F. (2015). An Updated Review on Cancer Risk Associated with Incretin Mimetics and Enhancers. *Journal of Environmental Science and Health, Part C: Environmental Carcinogenesis and Ecotoxicology Reviews*, *33*(April), 67–124.
- Vangoitsenhoven, R., Mathieu, C., & Van Der Schueren, B. (2012). GLP1 and cancer: Friend or foe? *Endocrine-Related Cancer*, *19*(5), 77–88.
- Villalba, A., Rodriguez-Fernandez, S., Perna-Barrull, D., Ampudia, R. M., Gomez-Muñoz, L., Pujol-Autonell, I., Aguilera, E., Coma, M., Cano-Sarabia, M., Vázquez, F., Verdaguer, J., & Vives-Pi, M. (2020). Repurposed Analog of GLP-1 Ameliorates Hyperglycemia in Type 1 Diabetic Mice Through Pancreatic Cell Reprogramming. *Frontiers in Endocrinology*, *11*(May), 1–14.
- Waget, A., Cabou, C., Masseboeuf, M., Cattan, P., Armanet, M., Karaca, M., Castel, J., Garret, C., Payros, G., Maida, A., Sulpice, T., Holst, J. J., Drucker, D. J., Magnan, C., & Burcelin, R. (2011). Physiological and pharmacological mechanisms through which the DPP-4 inhibitor sitagliptin regulates glycemia in mice. *Endocrinology*, *152*(8), 3018–3029.
- Xing, J., Li, Q., Zhang, S., Liu, H., Zhao, L., Cheng, H., Zhang, Y., Zhou, J., & Zhang, H. (2014). Identification of dipeptidyl peptidase IV inhibitors: Virtual screening, synthesis and biological evaluation. *Chemical Biology and Drug Design*, *84*(3), 364–377.
- Yabe, D., & Seino, Y. (2013). Incretin actions beyond the pancreas: Lessons from knockout mice. *Current Opinion in Pharmacology*, *13*(6), 946–953.
- Zhao, M., Chen, J., Yuan, Y., Zou, Z., Lai, X., Rahmani, D. M., Wang, F., Xi, Y., Huang, Q., & Bu, S. (2017). Dipeptidyl peptidase-4 inhibitors and cancer risk in patients with type 2 diabetes: A meta-Analysis of randomized clinical trials. *Scientific Reports*, *7*(1), 1–14.

CHAPTER

8

Vultures: Endangered Scavengers of the Thar Desert of Rajasthan

Ram Prakash Saran
Department of Zoology, Jai Narain Vyas University, Jodhpur (Rajasthan), India-342001
E-Mail: saranrp@live.com

ABSTRACT

Out of nine in subcontinent, there are seven species of vultures occurred in the Thar Desert of Rajasthan is of great significance. A systematic study carried out with the objective of demography, status, and nesting records of vulture species in the Thar Desert of Rajasthan. Potential records of nesting of resident vulture species also gathered at the right proportion. Major threats identified as habitat loss, collision from windmills, electrocution, trichobezoar are some of the likely reasons for declination of the vulture population.

Keywords: Demography, Nesting, Thar Desert, Vulture.

INTRODUCTION

Predators on top determine the health of the any ecosystem. The arid areas of the Thar desert of Rajasthan are richest from diversity perspective due to the good population of vultures as scavengers. They provide an important ecosystem service by feeding on carrion from deceased cattle. Nine species of Vulture show their presence in Indian subcontinent (Ali & Ripley 1987). Out of these nine species, Rajasthan represents four species of *Gyps* vulture *viz. Gyps bengalensis, Gyps indicus, Sarcogyps calvus* and *Gyps tenuirostris* listed as critically endangered in the IUCN Red data book (IUCN, 2021). This genus represents eight species worldwide; among them, four are resident species in Asia (*G. bengalensis, G. indicus, G. himalayensis*, and *G. tenuirostris*), three are found primarily in Africa (*G. africanus, G. coprotheres*, and *G. rueppellii*), and one breed in Eurasia but migrates into Africa and South Asia (*G. fulvus*) (IUCN, 2021). Other than this, *Sarcogyps calvus* and *Neophron percnopterus* are two other vulture species present in India and Asian boundaries. Vulture species have legal protection in India and are listed in Sch I (Part III) of IWPA (Indian Wildlife Protection Act, 1972). Seven species of vulture show their presence in Rajasthan state. Four of them are resident (*Gyps bengalensis, Gyps indicus, Sarcogyps calvus,* and *Neophron percnopterus*) while three are winter migrant (*Gyps fulvus, Gyps himalayensis,* and *Aegypius monachus*) (Saran, 2017).

There has been a drastic decline in the vulture population in Indian sub-continent over last two decades (Prakash, 1996, 1999; Prakash and Rahmani, 1999; Prakash *et al.* 2003; Chhangani, 2015; Gilbert *et al.* 2006; Purohit & Saran, 2013). From the early nineties, severe fall of *Gyps* population across the Indian continent was observed (Gilbert *et al.*, 2002; 2006). Three vulture species (Oriental White-backed *Gyps bengalensis*, Long-billed

G. indicus, and Slender-billed *G. tenuirostris)* population reduces by more than 97% and assigned as "critically endangered" in the entire south Asian (Prakash *et al.*, 2003).

The Thar Desert is a biodiversity-rich, open, semi-arid agro-grassland habitat due to wildlife and adequate domestic livestock. This region characterizes xeric vegetation and resources but imperfect human contravention. Very few studies commenced in the region in context to these highly delicate component of scavengers in the Thar area. Since 1996, the breeding ecology and population of the vultures studied the Thar desert were examined by monitoring their nesting sites, making censuses, recording predation, observing inter-species interaction and locating their seasonal migration (Chhangani and Mohnot, 2004).

In this study, we gathered the data on recent status, records of residents and migratory vulture species. Additionally, we describe previously unreported colonies of White-backed Vultures, Red-headed vultures, Long-billed Vulture, and Egyptian vultures in the Thar area

Material and Method

Thar Desert of Rajasthan covers 12 districts of Rajasthan namely Barmer, Bikaner, Churu, Ganganagar, Hanumangarh, Jaisalmer, Jhunjhunu, Jodhpur, Jalor, Nagaur, Pali and Sikar ranging from 24°37' to 30°10'48" north latitude and 69°29' to 76°05'33" east longitude (Fig. 1). The study area involves semi-arid and arid desert landscape with dry condition. The climate is characterized by low rainfall with erratic distribution, excesses of daily and annual temperatures, low humidity and high wind velocity.

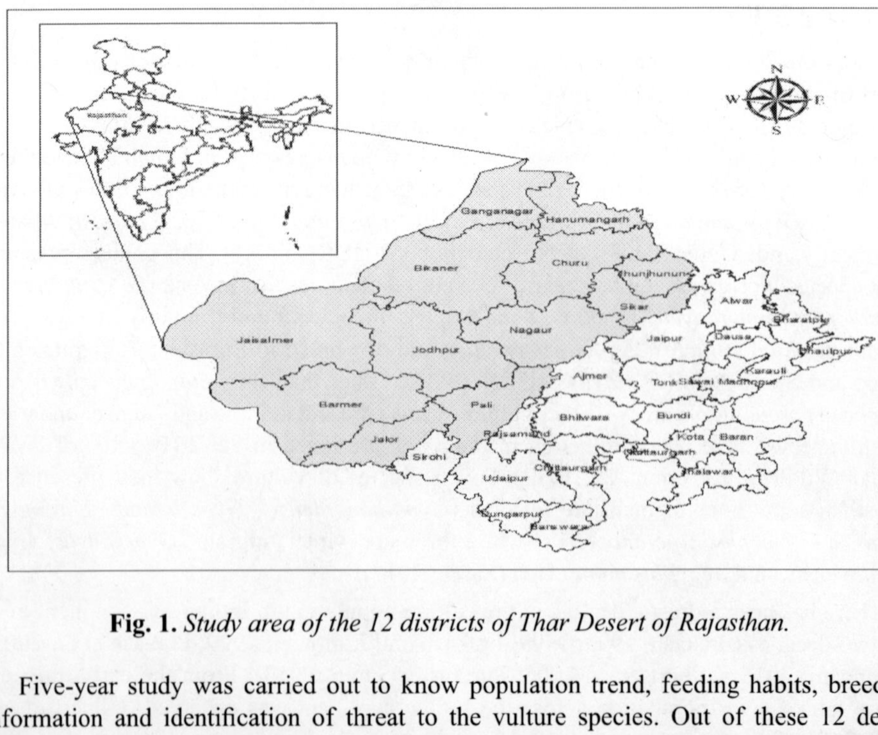

Fig. 1. *Study area of the 12 districts of Thar Desert of Rajasthan.*

Five-year study was carried out to know population trend, feeding habits, breeding information and identification of threat to the vulture species. Out of these 12 desert districts, special emphasis was paid to Bikaner, Barmer, Jaisalmer and Jodhpur districts. Opportunistic observations were also made as and when possible. The information about various roosting, nesting and feeding sites were used for study sites. Information was gathered from local people, surveying around water bodies, high rock cliffs, area of high tension electricity lines, carcass dumping grounds, high trees and information from livestock dwellers. During the survey, nesting vultures was thoroughly searched by scanning cliffs and nesting trees in open areas.

In order to minimize disturbance most observations were made with the help of field binoculars. In case of absence of species, nest was observed closely to find out nesting activity. A GPS device was used to take note of coordinates of the localities/place. Identification of the species was done with the help of bird field guide and key identification characters like body color pattern, upper and under wing color patterns, bill shape and thickness, shape of tail. The population was counted by using different methods like head count, photographic count and video counts. Feeding behavior and assessment of threats were directly observed in ~~at~~ the field and observations were recorded.

RESULTS

Out of nine in Indian sub-continent, seven species of vultures are present in Rajasthan. Major characters for identification of different species are body size, shape and size of bill, scarf, shape of head and body color on dorsal and ventral wing color and patterns.

Species Account

1. Indian Vulture (*Gyps indicus*)

   Body Size: Medium size

   Bill: Upper part pale white lower part black

   Head and Neck: Head black in color, neck thin covered with rufus hair

   Scarf: white in color

   Feather, Plumage: Feather black-brown and of feather is creamish.

   Feather Strata: Strata three leg plumage is white in color.

2. White Backed Vulture (*Gyps bengalensis*)

   Body size: Smaller in size

   Bill: Upper part: creamy white lower part: black

   Head and Neck: Neck without hair Grey-black

   Scarf: white in color

   Feather, Plumage: Back: White Feather: black, Lower side of feather have white band.

   Feather strata: Strata not well developed but between feathers there is a gap.

3. Egyptian Vulture (*Neophron percnopterus*)

   Body size: Small

   Bill: Narrow with black tip

   Head: Small and pointed

   Scarf: Yellowish in color

   Feather, Plumage: deep-fingered black flight feathers contrasting with white inner feathers. Juveniles are dirty brown with bare gray faces.

4. Red-Headed Vulture (*Sarcogyps calvus*)
   Body Size: Small to medium
   Bill: Red with Black tip
   Head and Neck: head hairless, Two Broad skin folds (Lappets)
   Scarf: white deep prominent
   Feather, Plumage: Feather is completely black in color from upper and dull at lower side
   Feather strata: Long black plumage up to the half of legs.
5. Eurasian Griffon (*Gyps fulvus*)
   Body size: Large
   Bill: Smaller then *G. himalayensis*
   Head and Neck: Upper portion is pale crimish in color, lower portion black in color
   Scarf: Pale white and neck covered with Rufus white hair.
   Feather, Plumage: Pale white and scarf hair short then *G. himalayensis*.
   Feather strata: Roufous brown plumage, short in size, No gap between plumage, tip end of plumage is round.
6. Himalayan Griffon (*Gyps himalyensis*)
   Body size: Large
   Bill: Tip pale and remaining black
   Head and Neck: Dark in color (black and brown)
   Scarf: Long, bright white at root and at tip brown in color
   Feather, Plumage: Tail wing long, plumage thin and long, middle row of single plumage is white in color
   Feather strata: Feather with 4 strata and tail and leg plumage is long.
7. Cinereous Vulture (*Aegypus monachus*)
   Body size: Large
   Bill: Large
   Head and Neck: Tip black and remaining pale white in color
   Scarf: Neck short, thick and covered with black plumage
   Feather, Plumage: Scarf covered with long hair; black color
   Feather strata: Feather completely creamy or brown in color

Population Trend

This study shows results from five-year population monitoring which was carried out to know current population trend of resident species of vultures from 2016-2020. Study was conducted at Jodhpur district. The important areas from viewpoint of vulture population were surveyed round the year and opportunistic observations were also taken into account. The population trend of the migratory and resident species is as under.

Year	Migratory species			Resident species		
	G. himalayensis	***G. fulvus***	***A. monachus***	***G. indicus***	***G. bengalensis***	***N. percnopterus***
2016	96±12	223±28	62±22	24±12	0±1.27	1084±22
2017	92±10	287±45	56±19	18±18	0±2.27	1127±27
2018	88±14	310±21	51±22	22±19	0±2.22	1258±29
2019	81±22	328±26	48±16	19±13	0±2.30	1267±33
2020*	43±19	157±34	28±33	10±4	0±1.70	957±30

* Relatively low mean is subjected to sampling gaps due to corona pandemic during the period.

Among migratory vultures, overall trend of *G. fulvus* seems to increasing and in case of resident species *N. percnopterus* it is also seems to survive very well in study area. *N. percnopterus* is the most commonly seen around all sites and among migratory species *G. fulvus* is the most abundant and noticeable at feeding areas. *G. himalayensis, A. monachus, G. indicus, G. bengalensis* showing a downtrend in general which is subject to be monitored further for the conservation point of view.

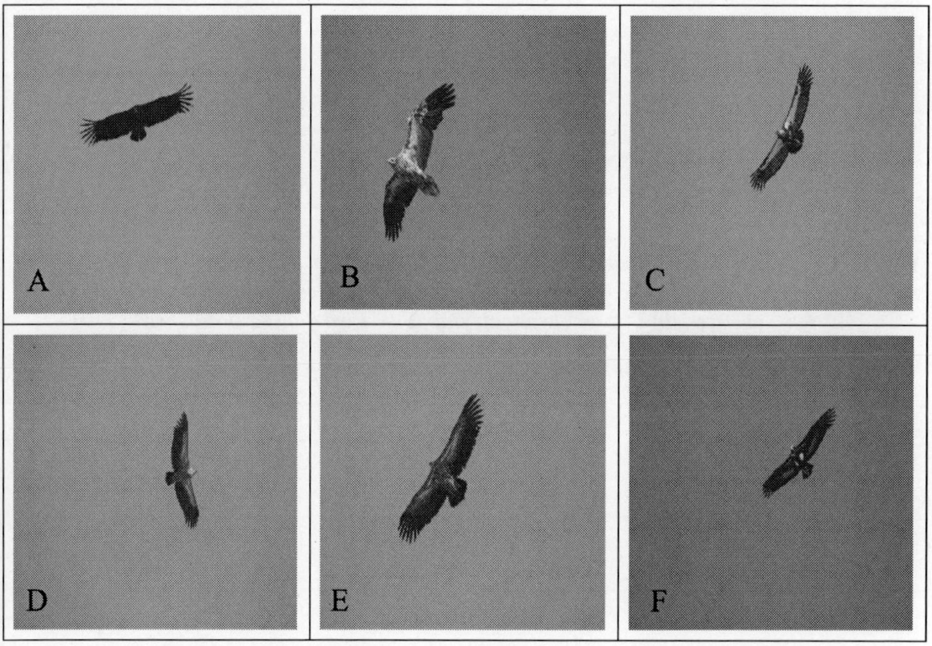

Fig. 2: *Vulture species at study area: A. A. monachus, B. N. percnopterus C. G. bengalensis D. G. fulvus, E. G. himalayensis, F. S. calvus*

Fig. 3: *A Gyps indicus nesting at Eklingnath Ji site; B. Neophron percnopterus at Balicha Dump site, Udaipur; C: Gyps fulvus, Neophron percnopterus and Aegipus monachus at Jorbeer, Bikaner (Camera trap); D: Sarcogyps calvas at Bhadariya Rai site, Jaisalmer.*

Table 1: *IUCN Redlist status of Nine species of Vultures present in India*

S.No.	Species	Common name	IUCN Redlist category
1.	*Aegypius monachus**	Cinereous Vulture	NT
2.	*Gypaetus barbatus*	Bearded Vulture	NT
3.	*Gyps bengalensis*#*	White-rumped Vulture, White-backed Vulture	CR
4.	*Gyps indicus*#*	Indian Vulture, Long billed Vulture	CR
5.	*Gyps himalayensis**	Himalayan Griffon Vulture	NT
6.	*Gyps fulvus**	Griffon Vulture	LC
7.	*Gyps tinuirostris*	Slender-billed Vulture	CR
8.	*Sarcogyps calvas*#*	Red-headed Vulture	CR
9.	*Neophron percnopterus*#*	Egyptian Vulture	EN

* Species present in Rajasthan, India; # Resident species of Rajasthan.

Source: IUCN, 2021

Food and Feeding Behavior

Vultures provide important ecosystem service by eating dead and decaying carcasses and are called as natures scavengers. Although vultures tend to soar high in search of food but municipal dump yard are proved to be their favorite places as they provide food security to vultures. All municipal areas have their specific dump site for disposal of dead animal carcasses. Unlike natural places, supply of dead animals is almost regular to these places. At these feeding sites, ground staff skin dead animal and leave rest of carcass for vultures and other animal consumption. Keru is a main dump site located approximately 3 km away from the Jodhpur city. Here dead animals are dumped for disposal from Jodhpur city, Nagore, villages and other areas around this dump site. Along with other domestic animals, majority of carcass dumped here were of cows.

Fig. 4: *A-Eurasian griffon dominating other species while feeding vigorously; B- Waterpoint at Jorbeer utilized by vultures after feeding.; C- Vultures basking after arrival at feeding site; D- Communal feeding of vultures*

Vultures show a communal feeding pattern, where they feed with other species of mammals (*Canis familiaris, Canis aureas*) and birds (*Corvus corax, Dicrurus macrocercus, Merops philippinus, Carvus splendens, Strunus roseus*) at the feeding site. A specific sequence of vulture feeding was observed at the feeding site. Opening of outer skin layer take about thirty minutes after that *Gyps fulvus* and *Gyps himalayensis*, *Aegypus monachus*, *Neophron percnopterus* start feeding on carcass. During this feeding *Gyps fulvus* and *Gyps himalayensis* dominate over *Aegypus monachus*, *Neophron percnopterus*. *Gyps fulvus* mainly consume soft tissues of carcass like heart, kidney, stomach and liver. After feeding and dominating on other species *Gyps fulvus* and *Gyps himalayensis* go to nearby field area while remaining species take opportunity of feeding. After feeding till late afternoon *Gyps* species leave the feeding area and visit waterpoints available nearby.

Threat for Vulture Populations

The population decrease in this area is due to habitat alteration, food scarcity deforestation, feral dog hunting, electrocution, wind will tower establishments etc., are noted in the study area.

We have observed mortality cases of a Himalayan griffon trapped in a *Prosopis cineraria* tree, A Cinereous vulture was injured due to collision wind mill near Sam. Several times road crushed Egyptian vulture were noted along the Khuhri –Myajlar road in difference visits during the study, feral dog hunting in Keru dumping Jodhpur, Eurasian griffon at Lathi, electrocuted Egyptian vulture in Jodhpur, trichobezoar also identified in Jodhpur in a case of Eurasian griffon mortality.

In the year 2008 we found that carcass disposal plant pose threat to existence of vultures as the processing directly caused food scarcity while disposing carcass by incineration. Electrocution from high tension lines was found lethal and 17 deaths of five species of vultures were reported during the year 2007-2008.

At Jorbeer conservation reserve although food supply is regular as this is the major carcass dumping station in the Bikaner but feral dogs are major problem. It is already reported that feral dogs compete for feeding with vultures by barking and chasing them away to get opportunity for feeding on carcasses (Saran and Purohit, 2014). They not only attack feeding vultures but also attack the visitors.

Fig. 4: *A: Feral dogs and wild boar feeding at Jorbeer site, Bikaner after sunset (Camera trap); B: Feral dogs actively feeding at day time Jorbeer site, Bikaner (Camera trap).*

DISCUSSION

During the mid of 1980 and 1990 vulture population at Keoladeo National Park, Bharatpur was reportedly declined (Prakash, 1999). Survey carried out by BNHS during 1991 to 1993 and 2000 reported sharp decline in the population of Oriental white backed vulture and long billed vultures nationwide (Samant, 1995; Prakash *et al.* 2003). A sharp decline of 92% was observed in Oriental white backed vulture and long billed vultures. A nine-year study carried out from 2007-2015 reported a decline in the population of migratory species at Jodhpur district of Rajasthan. This study suggested that this decline of population may be due to climate change and temperature fluctuation during these periods (Saran, 2017). During this study we found that population trend in Jodhpur area. Recent studies on vultures of the Thar desert found increase in *N. percnopterus* and Eurasian griffon (Saran *et al.*, 2020). Similar results were obtained during this study.

Vultures are nature's scavengers which consume dead and decaying animals and help us to keep our surrounding clean. They show a peculiar behavior and hierarchy during feeding. Food availability is important for vulture conservation. Dump sites provide safe heaven to vultures as these are the places has abundant and regular food supply.

The Arid region of the Thar experience life-threatening environmental circumstances, not much favorable for civilization. But during the last 20 years, there was an increase in human population, which increased demands for food supply and land for anthropogenic activities. As a result of this demand, agricultural and residential land expended which ultimately led to degradation or destruction of the habitats and in serious concerns. Shrinking habitats disrupt animal movement and its foraging activities. Highly fragmented area cause disturbance for species that are sensible for human contact. Due to the increase in development activity a raise in mining activities is (particularly stone mining) also seen. Stone mining works require use of explosive materials for blast and heavy machines to dig out stone slabs, which create loud sound and cause disturbance to the nesting sites on hill cliffs and cause adverse effect. This type of habitat disturbance was seen in Ekling nath Ji site, Udaipur.

That reduction in the population of migratory species was due to a number of factors. The Thar Desert provided arid and semi-arid biogeography with highly specific traditional agro-pastoral livelihoods in a meaningful manner. Vulture species and status composition at the Thar landscape given utmost attention in this study.

Declining population trend is a matter of great concern. Various threats were reported which could be responsible for declining trend of vulture population. Use of Diclofenac for cattle was found to be important factor. Electrocution is also reported a cause of Vultures and other raptors deaths. Being a large size bird, Vultures sometimes come in contact with high voltage electricity distribution lines.

Poisoning from diclofenac drug was found to be a major threat to Vulture species. This drug was responsible for kidney failure in vulture species (Oaks *et al.*, 2004). A study done on white backed vultures at various sites in Indian subcontinent concluded that diclofenac amount found in tissues of ungulate carcass was sufficient to cause vulture population decline of white backed vultures (Green *et al.*, 2007). After ban on diclofenac in 2006, a study was conducted by Cuthbert *et al.*, 2016 in nine states of India. They confirmed presence of visceral gout and diclofenac in *Gyps bengalensis, Gyps indicus and Gyps himalayensis*. Nimeuslide and Meloxicam were also found in the five and two vultures respectively.

Feral dogs hunting, electrocution, and trichobezoar are reasons came out for vulture death in the Thar region (Saran, 2017; Saran & Purohit, 2012, 2016). This study finds vulture's mortality in various ways, such as Himalayan griffon trapped in a Khejri tree at Lathi, A Cinereous vulture injured due to wind turbines near Sam. Several times road crushed Egyptian vulture noted along the Khuhri –Myajlar road in different visits during the study, feral dog hunting in Keru dumping, Jodhpur, Eurasian griffon at Lathi, electrocuted Egyptian vulture in Jodhpur and trichobezoar in Eurasian griffon in Jodhpur. Perch site availability is also a matter of concern for vulture conservation. Vultures are birds with relatively large body size and weight. They tend to run for few moments and then start a low flight and start soaring high. Sometime during their low flight, they become victim of speeding vehicles if they are nearby any highway. Alike case of Great Indian bustard, Vultures also collied with wind turbines and fall dead.

At Jodhpur and Jorbeer site feral dogs are major problem which compete with vultures for feeding. These dogs remain active during day and even late after sunset. These are also a problem for visitors. Along with dogs, wild boar also feed on remains of dead animals. Feral dogs are not only problem in Jorbeer, but also at several other places like Balicha dump site, Keru dump site, Bhadariya Rai site. Proper management strategies with the information present today should be utilized for conservation of vultures.

ACKNOWLEDGMENT

Author is thankful to DST-SERB for providing financial support for this research work vide grant no. (CRG/2019/002090).

REFERENCES

- Ali, S. and Ripley, S. D. (1987). Compact handbook of the birds of India and Pakistan together with those of Bangladesh, Nepal, Bhutan and Sri Lanka. 2nd ed. Delhi: *Oxford University Press*.
- BirdLife International. 2021. *Aegypius monachus. The IUCN Red List of Threatened Species* 2018: e.T22695231A131935194. https://dx.doi.org/10.2305/IUCN.UK.2018-2.RLTS.T22695231A131935194.en. Downloaded on 11 September 2021.
- Chhangani, A.K. and Mohnot, S.M. (2004): Is diclofenac the only cause of Vultures decline? *Current Science*. 87(11): 1496-1497.
- Chhangani, A.K. (2015): Population ecology of Vultures in the western Rajasthan, India. *Indian forester*. 131 (10):1373-1382.

- Gilbert, M., Virani, M.Z., Watson, R.T., Oaks, J.L., Benson, P.C., Khan, A.A., Ahmed, S., Chaudhry, J., Arshad, M., Mahmood, S. and Shah, Q.A. (2002): Breeding and mortality of oriental white-backed vulture *Gyps bengalensis* in Punjab Province, Pakistan. *Bird Conservation International*. 12: 311-326.
- Gilbert, M., Watson, R.T., Virani, M.Z., Oaks, J.L., Ahmed, S., Chaudhary, M.J.I., Arshad, M., Mahmood, S., Ali, A. and Khan, A.A. (2006): Rapid population declines and mortality clusters in three Oriental white-backed vulture *Gyps bengalensis* colonies in Pakistan due to diclofenac poisoning. *Oryx*. 40: 388–399.
- Green R. E., Taggart MA, Senacha, K.R., Raghavan B, Pain DJ, Jhala Y, *et al.* (2007) Rate of Decline of the Oriental White-Backed Vulture Population in India Estimated from a Survey of Diclofenac Residues in Carcasses of Ungulates. *PLoS ONE* 2(8): e686. https://doi.org/10.1371/journal.pone.0000686
- Oaks JL, Gilbert M, Virani MZ, Watson RT, Meteyer CU, *et al.* (2004) Diclofenac residues as the cause of vulture population declines in Pakistan. *Nature* 427: 630–633.
- Prakash, V. and Rahmani, A. R. (1999): Notes about the decline of Indian Vultures, with particular reference to Keoladeo National Park, *Vulture's News*, 41. pp. 6-13.
- Prakash, V. (1996): Status of vultures in Keoladeo National Park, Bharatpur, Rajasthan, with special reference to population crash in *Gyps* species. *Journal of Bombay Natural History Society*. 96 (3): 365-378.
- Prakash, V. (1999): Status of Vulture in Keoladeo National Park, Bharatpur, Rajasthan, with special reference to population crash in *Gyps* species, *Journal of the Bombay Natural History Society*, 96. pp. 365-368.
- Prakash, V.D., Pain, J., Chunningham, A.A., Donald, P.F., Prakash, N., Verma, A., Gargi, R., Kumar, S.S. and Rahmani, A.R. (2003): Catastrophic collapse of Indian white backed *Gyps bengalensis* and long billed *Gyps indicus* Vulture population, *Biological Conservation*, 109. pp. 381-390.
- Purohit, A. and Saran, R. (2013): Population Status and Feeding Behavior of Cinereous Vulture (*Aegypus monachus*): Dynamics and Implications for the Species Conservation in and Around Jodhpur. *World Journal of Zoology*.8(3):312-318.
- Saran, R. and Purohit, A. (2012): Eco transformation and electrocution. A major concern for the decline in Vulture population in and around Jodhpur. *International Journal of Conservation Science*. 3(2):111-118.
- Saran, R. and Purohit, A. (2014): Population status, nesting sites and seasonal fluctuation of Egyptian vultures (*Neophron percnopterus*): Dynamics and implications for the species conservation in and around Jodhpur. *International Journal of Biodiversity and Conservation*. 6(1):100-107
- Saran, R. and Purohit, A. (2016): Trichobezoar: A Possible Cause of Death in Eurasian Griffon *(Gyps fulvus)* - A Case study in Jodhpur, Rajasthan, India. *Zoo's Print*.31(12): 15-16.

- Saran, R. (2017): Population monitoring and annual population fluctuation of migratory and resident species of vultures in and around Jodhpur, Rajasthan. *Journal of Asia Pacific Biodivesity*.10(3):342-348.
- Saran, R., Kumawat, R., and Purohit, A. (2020): Recent trends of population and nesting of threatened vulture species in Thar Desert of Rajasthan, India. *Uttar pradesh journal of zoology*, 41(18), 44-53.

CHAPTER

9

Snakes: Annotated Checklist from Thar Desert of Rajasthan

Rakesh Kumawat and Ashok Purohit
Department of Zoology, Jai NarainVyas University, Jodhpur (Rajasthan), India-342001
E-Mail: rakeshophia@gmail.com

ABSTRACT

Ophiological diversity was assessed in the arid and semi-arid regions of the Thar Desert in Rajasthan. The need of this research came out after the significant implications on the local topography of westerns districts of Rajasthan after the Indira Gandhi Nahar Pariyojna canal expansion for three decades. Present research paper updates the status of snakes in the outline of diversity and occurrence in the golden triangle of the Thar Desert of Rajasthan state (Jodhpur, Jaisalmer, and Bikaner). The findings are based on the four years of extensive field study through an active search, night drives, scale counting, and topographical database. The study resulted in working out, 22 species and one subspecies.

Key Words: Diversity, Snake, Thar Desert.

INTRODUCTION

Owing to the unique geographical and geological characters, the Thar Desert of Rajasthan has a unique diversity of reptiles in general and snakes in particular. The faunal composition in the Thar region holds a wide variety of habitat and high diversity, due to juxtaposition of Palaearctic, Oriental, and Saharan elements (Blanford, 1901; Minton, 1966).

The western region of Thar fascinates numerous workers to study serpent in terms of diversity, taxonomy, morphology, natural history and distribution. But still uncommon, promising and discrete records were obtainable for the ophidea group (Das, 1997; Sharma, 2003, 2007; 2013; Sharma & Rathore, 2004). In western Rajasthan, 16 snake species reported in a thorough study of Sinha & Sharma (2008). Though, Whitaker & Captain (2004) provided detailed information, the present study attempts to deliver a new data set the modification and change of landuse in the ecosystem of the desert after the introduction of Indira Gandhi Nahar Canal and its impact in the region. Recent study of rescue and rehabilitation in two of the desert urban areas (Jodhpur and Jaisalmer) accounted 14 species of snakes (Husain & Purohit, 2019). Despite this, present study reports 22 species and one sub-species of seven families from Jodhpur, Jaisalmer and Bikaner districts (Kumawat & Purohit, 2020). The occurrence, distribution patterns, and population dynamics studies of known species of snakes still waited from these selected areas. Hence, the present work anticipates various morphological and distributional details, which have usually rummage-sale to define these taxa.

Though the last three decades, the semi-arid and arid terrain has endured complete eco-transformation but it also has an increase in biotic burden interrelated with the insubstantial environmental entities to create frightening environmental glitches. The present study records a new data set from the desert's changing ecosystem after escalating impact of the Indira Gandhi Nahar Canal. As, the knowledge of the diversity and distribution of the snake in the Thar Desert of Rajasthan is somewhat scanty, and the snake dwelling in this region had no clear picture/diversity representation thus, indispensable data on life history characteristics for understanding the species ecology and comprehensive distribution data for biogeographic studies and conservation assessments are still missing for an extensive proportion of important diverse areas of the Thar Desert of Rajasthan.

Material and Methods

Fieldwork was concentrated principally in three districts where a variety of microhabitat available for the snakes. The golden triangle includes Jodhpur 26.2800° N, 73.0200° E, Jaisalmer 26.9200° N, 70.9000° E, and Bikaner 28.0167° N, 73.3119° E districts that represent unique habitat characteristics. Surveys were plotted over the entire year from September 2015 to August 2019 for the purpose of snake search in Golden triangle (Bikaner, Jodhpur, and Jaisalmer district). Visual and opportunistic encounter methods were used to record the snakes. Different characteristics of habitat searched for snake viz. Grassland, Stabilized dunes, Barren dunes, Human habitation, and IGNP water canal area. The Snakes were photographed, examined, and released at the previously desert habitat to around (Map 1).

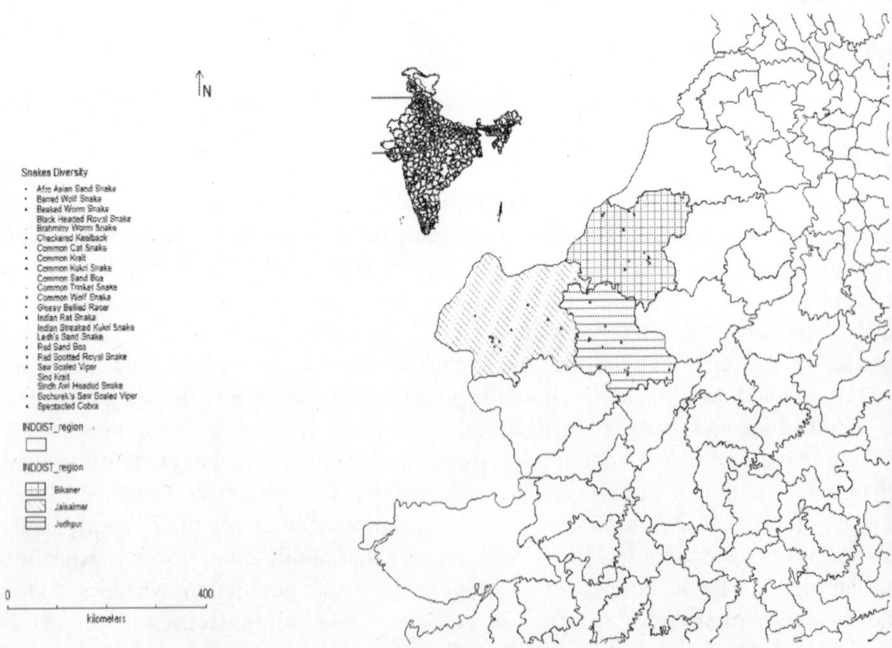

Map. 1. *Map of snakes species recorded from Jodhpur, Jaisalmer and Bikaner district of Thar Desert of Rajasthan.*

Species Account

Family: Typhlopidae (Merren, 1820)

Brahmin Worm Snake *Indotyphlops braminus* (Daudin, 1803)

Dorsal coloration of this snake is light brown, maroon to black; rostral, cloaca, and tail tip is a bit whitish; ventrum is light brown; smallest snake species in India; The total length recorded of an individual is 210 mm., fifteen individuals recorded in the study period. **Habitat:** This snake prefers humid places, pots, and gardens. In the rainy season, it is frequently observed near the water ditches and human habitation. **Activity**: Crepuscular.

Beaked Worm Snake *Grypotyphlops acutus* (Duméril & Bibron, 1844)

Dorsal coloration is brown; the ventrum is light brown. A most giant amongst worm snakes; long, cylindrical body; A beak-like projection on the snout with spiky flat border; rostral shield very massive and extended to the eyes; nostrils are poor and diminutive. Total length recorded 520 mm in an individual. **Habitat**: Similar to BWS but comparatively rare. It also prefers humid places equipped with subterranean lifestyle. Termites, ants, and other insects are the prey of this snake. **Activity**: Crepuscular.

Family: Erycidae (Grey, 1825)

Common Sand Boa *Eryx conicus* (Reynolds & Henderson, 2018)

Dorsal coloration is greyish, mixed with a dorsal chain of dark-brown, black-edged amalgamated spots; and laterally irregular and scattered; a dark band at the rear of the eye; the outer scale-rows are with spotted grey flecks in some cases; ventrum is whitish. Stout bodied, round and very dumpy tail; the head is somewhat distinctive from the neck and covered with small scale but nasal, and inter-nasals are prominent; Rostral is wider than longer; nostril is a slit-like in between nasals and inter-nasals. Most extended individuals recorded 780 mm; tail length is 93 mm. **Habitat:** Rocky terrain, Agricultural field, and open areas. It is a lethargic snake, strong burrower; Aggressive, and often bites when careless handling. This snake is viviparous. The food of this species includes rodents, birds, and reptiles, suffocate the prey by constriction. **Activity:** Diurnal; also recorded while basking at early morning in the winter season.

Red Sand Boa *Eryx johnii* (Russell, 1801)

Dorsal coloration is dark red to maroon, In juveniles homogeneously chocolate colored to dark brown, pale mottling flanks with hazy and light to darker slanting bands tail rings, which are more distinct on the tail region; ventrum is whitish with dark brown spots. Robust and cylindrical bodied; tapered abruptly tail after cloaca; head not distinct from the neck. Rostral wider than long; the Snout is moderately broad and wedge-shaped; nostril is slit-like and located between the two enlarged nasal scales. The snout-Vent ratio recorded in an individual is 268 mm; Total length to Tail length ratio is 9. **Habitat**: Red sand boa is submissive, mild, lethargic, and prefers to live in arid areas. The population observed in all possible habitats in the study area. This snake is viviparous. Its food comprises small mammals, birds, and insects, and the prey killed by constriction. **Activity**: Nocturnal.

Family: Colubridae (Oppel, 1811)
Subfamily-Colubrinae

Common Trinket Snake *Coelognathus helena helena* (Daudin, 1803)

Dorsal coloration is light brown, olive green with brown cross-bars. At anterior, two broad dark stripes on the lateral side and a black vertical streak are below the eye with a leaning one behind it; the ventrum is unvaryingly yellowish. Body laterally compressed; long Snout; head distinct; tail is long. Loreal length is longer than its height; Pre-frontals are longer than inter-nasal shields. Pre-ocular is 1; post-ocular is 2. Pre-frontals are larger than inter-nasals. Frontal is broader at anterior and pointed behind. Snout to vent length recorded is 175-360 cm; maximum total length recorded is 360 cm. **Habitat:** This snake is an alert and pugnacious species; it occupies the plantation and nursery areas of IGNPC and nearby urban areas. **Activity**: Nocturnal, also rescues in the day time in the nearby area of Canal.

Indian Rat Snake *Ptyas mucosa* (Linnaeus, 1758)

Dorsal coloration is olive green, dark olive-brown; many subsequent scales are with black edges, and anterior scales are with yellow or light tan edges; Supra-labials and Infra-labials and lateral head shields noticeably edged with black; blackish boundaries of the ventrals and sub-caudal scales; ventrum is muddy. The body is cylindrical, tapering tail and muscular, head distinct. The snout pointed; rostral is wider than its stature; nostrils located between nasals and 1st supra-labial; loreal is 3. Pre-ocular is 2 (upper one is larger); Post-ocular is 2. Frontal is broader at anterior and pointed behind. Temporals are two on either side. Snout- vent length ranges in between is 630-1300cm. **Habitat**: It is known to occur in a variety of habitats. Here, one individual observed from Nehru garden in Jodhpur urban area and another from the IGNPC area of Poogal in Bikaner district. **Activity:** Diurnal.

Glossy-Bellied Racer *Platyceps ventromaculatus* (Gray, 1834)

Dorsal coloration is light greyish to sandy with black cross-bars and rhomboidal spots, with brown or grey scales edged with black and a broken series of tiny spots and a diminutive vertebral stripe on neck; a slanted black stripe below the eye and on temporal; ventrum is whitish. The body is slender, cylindrical, and tapering; slightly distinct head. Rostral is higher than wide; nostril is prominent and pierced in 2 nasal shields; loreal is between nasal and pre-ocular. 2 pre-ocular (the upper one is ~ 4 times larger); post-ocular 2 of almost equal size. Pre-frontals are slightly larger than inter-nasals; anterior frontal edges touched the upper pre-ocular on either side laterally. Parietals are regular. The data recorded as SVL (180-770 mm), Tail (56-220 mm). **Habitat:** The snake shares stabilised dunes, rocky terrain, urban areas, agricultural field, grasslands, and IGNPC areas. **Activity** Diurnal. This snake is noted all over the year.

Black-Headed Royal Snake *Spalerosophis atriceps* (Fischer, 1885)

This snake's dorsal color is rough blackish, with yellow spots scattered all over the body. The head partially or entirely black or with a deep red shade; The yellow color is restricted to single scales or thickened into large rhomboidal dorsal spot; Adult snakes exhibit plenty of variation and individual with yellow, orange, pink, and with extremely variable flecks and blotch presents on the whole body. Juveniles with grey to light brown dorsum with blackish rhomboid oval bands; Ventrum is mottled reddish pink with the ashy background. Large bodied snakes; the head is relatively distinct; shortened Snout. Rostral is higher than wide; nostrils pierced in 2 nasal shields; loreal are 2, rarely 3. Oculars are 9-10. Frontal is regular; pre-frontal divided into 7-9 asymmetrical scales. The data recorded as the snout-vent length ranges between 430-1570 mm. **Habitat:**Maximum of this species observed of black morph in Bikaner urban area. Inhabit open scrubby areas and live in burrows, rock crevices, abandoned in old. Grassland, urban area, the agricultural field, stabilized dune, rocky terrain, and IGNPC area were the primary habitat recorded. **Activity:**Diurnal.

Red-Spotted Royal Snake *Spalerosophis arenarius* (Boulenger, 1890)

Dorsal coloration is fawn caramel with seven rows ruddy markings; broader at mid-body; proportioned markings on between the eyes; First arched band on head continues as a slanting strip behind the eye to mouth; Ventrum is glossy. Medium to the long cylindrical bodied snake; the head is distinct from the neck. Snout-vent length is in between 530-1300 mm. **Habitat**: This snake usually prefers rat holes in open scrubby grassland and agricultural area. In contrast to BHRS, this snake is timid and not observed from the study area's urban areas. The habitat preference of this snake is grassland, stabilized dunes, agricultural areas, and IGNPC area.**Activity**:Nocturnal.

Sindh Awl-Headed Snake *Lytorhynchus paradoxus* (Gunther, 1875)

Dorsal coloration is greyish white to pale brown with a median series of X or H shaped brown to sooty blotches separated by cream-colored interspaces and a pale tan lateral series of smaller spots. On the head, an outsized rhomboidal brown spot present behind the parietals and a brown strip behind the eye; the ventrum is glossy white. It is a short-sized snake with a cylindrical body appearance, wedge-shaped projecting Snout, head not distinct from the neck. Rostral is greatly prognostic, curved in low-grade; Nostril is a slanting slit. Pre-ocular is 2; post-ocular is 2; 1 pre-sub ocular is sorting out the 4th supra-labial from the eye. Snout –vent length is in the range of 213-405 mm. **Habitat:** This snake spent day time under the soft layer of loose soil. They have a very narrow range of habitat preferences in dunes, both stabilized and barren. **Activity**: Crepuscular and Nocturnal.

Indian Streaked Kukri snake *Oligodon taeniolatus*(Jerdon, 1853)

Dorsal coloration is light chocolate; blackish or brown; with narrow rounded cross bars; a dark band across the head upon eyes; nape with M-shaped dark neckline; ventrum is greyish. The body is slender; blunt Snout; somewhat different head. Rostral is higher than wider, extending later and break up the inter-nasal shields; nostrils placed between the nasal shields. Pre-ocular is 1; post-ocular are 2. Frontal and pre-frontals (2) are regular. Temporal is 1. The snout-vent length is 440mm; Tail length is 80mm. **Habitat:** The snake found in rocky areas and gardens upon bushes. **Activity:** Nocturnal.

Common Kukri Snake *Oligodon arnensis* (Shaw, 1802)

Dorsal coloration is golden-brown; discrete cross-bars on the whole body till tail point; head with three inverted V-shaped marks; ventrum is whitish with faded lateral spots. The body is slender. Loreal is in contact with the pre-frontal. Pre-ocular is 1; post-ocular are 2. Frontal and pre-frontals (2) are regular. Temporal is 1. Snout- vent length is 560mm; Tail length is 95mm. **Habitat:** This snake observed from the nursery areas of the IGNPC area. **Activity**: Diurnal.

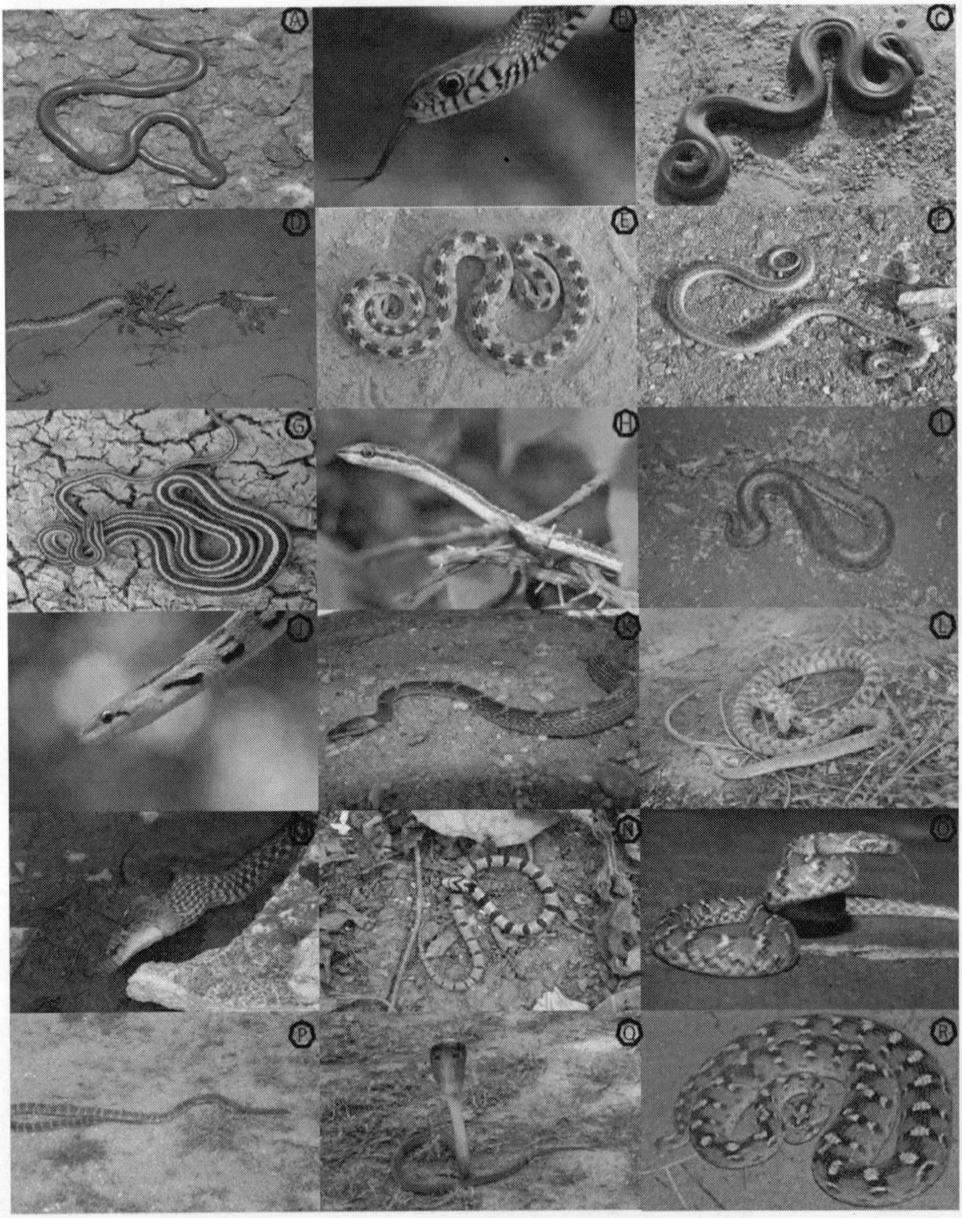

Fig. 1. *Various snake species: (A) Brahmin Worm Snake, (B) Rat Snake, (C) Red-Sand Boa, (D) Red-Spotted Royal Snake, (E) Sindh Awl-Headed Snake, (F) Russell's Kukri, (G) Leith's Sand Snake, (H) Afro-Asian Sand Snake, (I) Common Sand Boa, (J) Trinket Snake, (K) Common Wolf Snake, (L) Glossy-Bellied Racer, (M) Fresh water snake, (N) Common Kukri Snake, (O) Common Cat Snake, (P) Common Krait, (Q) Spectacled Cobra and (R) Saw Scaled Viper.*

Common Wolf Snake *Lycodon aulicus* (Linnaeus, 1754)

Dorsal coloration is brown with yellow cross-bars; V-shaped yellow band on neckline; the upper lip is white; ventrum is whitish—medium-sized extended body with dejected head. Nostril located between two nasals. Pre-ocular is 1; post-ocular are 2. Frontal is regular; pre-frontals are noticeably larger than inter-nasals. Temporals are 2+2 or 2+3. Snout –vent length is 523mm; Tail length is 99mm. **Habitat:** One of the excellent climbers and prefers to live in or around human habitations. **Activity:** Nocturnal.

Barred Wolf Snake *Lycodon striatus* (Shaw, 1802)

Dorsal coloration is dark-brown, with white cross-bars from head to tail; get more prominent on the lateral side; upper lips and ventral parts are white. This snake is a slim bodied; poorly distinct head. Rostral is wider than higher; loreal 1; touches the inter-nasal. Pre-ocular is 1; post-ocular are 2. -Frontal is regular, and pre-frontals are noticeably larger than inter-nasals. Temporals are 2+2 or 2+3. Snout – vent length is 380mm; Tail Length is 90mm. **Habitat:** Agriculture fields near to IGNP canal are. **Activity:** Nocturnal.

Common Cat Snake *Boiga trigonata* (Bechstein, 1802)

Dorsal coloration is pale brown, grey; two irregular series of large, elliptical, edged dark-brown spots; and two series of much smaller spots on the flanks; small spots on a mid-vertebra; two black longitudinal stripes upon the head; nape is with a longitudinal blotch; a stripe emerges from the eye and reach up to the angle of mouth; a series of brown spot on lateral sided of the body; ventrum is whitish or spotted. The body is a long, cylindrical, triangular head diverse from the neck. Rostral is blunt, wider than higher; Nostril between 2 nasals. Pre-ocular is 1. The frontal is relatively dumpy; pre-frontals are usual. Temporals are 2+3. Total length is – 340mm, the Tail length is 50 mm. **Habitat:** Cat snake is an arboreal species, prefers medium bushes, scrubby vegetation, and near humans. This snake occurs in all the represented habitats except barren dunes. **Activity:** Nocturnal.

Family: Natricidae (Oppel 1811)

Checkered Keelback *Fowlea piscator* (Schneider, 1799)

Dorsal coloration is yellow, olive, light green with five rows of small light black spots; bars which are generally narrow that interspaces are separating them; two oblique black streaks on the head; one below; second behind the eye. The black dorsal markings are arranged in five transverse series and form a chess-board like pattern; Ventrum is creamy white, with black spots on the laterally. The robust body; the head is elliptical, dejected, and blunt. Rostral is wider than higher; nostril is between nasals; loreal 1. Pre-ocular is 1; post-ocular are 3. Frontal, and pre-frontals are usual. Temporals are 2+2 or 2+3. Total length recorded is1350cm. In percentage of Total length, tail length is 4.28. **Habitat:**Habitat preferences observe of this snake is IGNP canal area, rocky terrain, and urban area. **Activity:** Diurnal.

Family:Psammophidae (Bourgeois, 1968)
Subfamily- Pammophiinae

Afro-Asian Sand Snake *Psammophis schokari* (Forskal, 1775)

Dorsal coloration is yellowish, grey, pale olive or reddish above; a dark streak on per side of the head, crossing through the eye; labials with dark dots; non-striped morph is without stripes, greyish; strips on had limited the head region; stripped form possesses four dark brown longitudinal stripes till the tail end; Ventrum is with dark dots and with 1 or 2 broken dark lines on each side. The body is slender, elongated, and cylindrical; head distinct from neck. Rostral broader than high; nostril is between nasals; loreal 1 stretched. Pre-ocular touches frontal and direct with the anterior edge of pre-frontal; Post-ocular is 2. The anterior end of the frontal is twice as broad as the middle; pre-frontals are 2. Parietals are asymmetrical and somewhat longer than frontal; Temporals is 2+2. Snout to vent length recorded between 340-986mm. In percentage to total length, the tail length recorded as 4.03-4.28. **Habitat:** This snake is a semi-arboreal inhabits in sandy desert with sparse vegetation areas; records also have been observed from the grassland and agricultural areas. **Activity:** Diurnal.

Leith's Sand Snake *Psammophis leithii* (Gunther, 1869)

Dorsal coloration is yellowish, grey, or pale orangey; lateral facet is much darker; two deep-brown stripes begins from inter-nasal along the whole length of the body and broader on the posterior side; head dorsally dark and possess middle, longitudinal stripe; a dark stripe passes through the eyes; labials are of cream colour; ventrum is uniformly yellowish. The slender extended body; the head is distinct from the neck. Inter-nasals are about half of the size of pre-frontals. Pre-ocular touches the frontal; post-ocular is 2. The anterior end of the frontal is twice as broad as the middle; pre-frontals are 2. Temporals are 1+2 or 2+2. Total length is observed 690mm, the Total length tail is 235mm. **Habitat**: Single record of this species observed in an agricultural area near a humid place. **Activity**: Diurnal.

Family: Elapidae (Boie, 1827)

Common Indian Krait *Bungarus caeruleus* (Schneider, 1801)

Dorsal coloration is black with a series of vertebral spots on the neck in juveniles; the body has narrow, transverse bands in pairs; upper labials are yellow or whitish; ventrum is pale whitish. The medium-sized cylindrical body; Head not distinct; No loreal scale. Pre-ocular is 1 of large-sized; post-ocular is 2 (small but same sized). Frontal is regular; pre-frontals are larger than supra-ocular and inter-nasals. Temporals are 1+2. Snout –vent length recorded is 790cm; in percentage to total length, tail length is 7.58. **Habitat**: Strongly snake confined to semi-arid areas, rocky areas, and agricultural areas with the urbanized settlement. **Activity:** Nocturnal.

Sindh Krait *Bungarus sindanus sindanus* (Boulenger, 1897)

Colour is black above, with slender white transverse bars; Whitish supralabials and ventrum. The body is solid and comparatively more significant than the Common Krait. Loreal is absent.). Frontal is regular; pre-frontals are larger than supra-ocular and inter-nasals. Supra-labials are 7/7, 3rd, and 4th touches the eye; infra-labials are 7/7. Big eyes are with a round pupil. Big eyes are with a round pupil. Fixed poison fangs; proteroglyphous; Snout –vent length recorded is 1090cm; Tail length is 110cm.**Habitat:** It confines Rocky terrain, an agricultural area, and humid areas of IGNPC. **Activity**: Nocturnal.

Spectacles Cobra *Naja naja* (Linnaeus, 1758)

Dorsal coloration is uniformly jet black or dark olive; spectacle mark on the dorsal side of hood prominent in juveniles and faded in adults and sometimes unclear; hood on ventral side posses dark transverse bars clear or dull sometimes; ventrum is pale gray. Ribs at the head region stretched out; neck prolonged into the hood. Nostril is prominent; no loreal. Frontal and Pre-frontals are usual. Temporals are 2+3 or 2+4. Supra-labials are 7/7, 3rd and 4th touches the eye, 3rd is largest also touches the nasal shield. Total length recorded as 760cm, tail length 210cm. **Habitat:** Cobra occurs in all the habitats except barren dunes. **Activity:** Diurnal but equally occurred in late nights.

Family: Viperidae (Oppel, 1811)

Indian Saw-Scaled Viper *Echis carinatus carinatus* (Schneider, 1801)

Dorsal coloration is from light buff or paler or olive-brown; the dorsal face of the head is with light arrowhead mark; dorsum is with whitish spots; a paler stripe begins from the eye and reaches up to the angle of mouth; labials are with dim brown spots with darker margins; lateral aspects are with narrow; undulating, white line; ventrum is whitish with rough dark gray stripes. Small bodied; the head is moderately distinct from the neck. Nostrils are minute and placed in dorsolateral place in a nasal shield.14 -17 oculars around the eye. Total length is 521mm; Tail length is 47mm. **Habitat:** Saw scaled viper is the most adaptable snake in the study area except in sandy, rocky, and semi-arid or arid environments with grass, herbs, and shrubs. **Activity:** Nocturnal.

Sochurek's Saw-Scaled Viper (*Echis carinatus sochureki* (Stemmler, 1969)

Dorsal coloration is similar to E. *carinatus carinatus*; the Arrow mark on the dorsal head is sometimes differing. The body is comparatively large. 17-21 ocular scale around the eyes; Supralabials separated with subocular series from the eye; sometimes, one pre-sub-ocular is present at the anterior side - twelve scales between the eyes at the dorsal surface. Total length is 330-364mm; Snout to vent length is 292- 321; Tail length is 38- 43. **Habitat:** Sochurek's Saw scale viper has occurred in the areas of the rocky and semi-arid or arid environment of Desert national park. **Activity:** Nocturnal.

CONCLUSION

The present study exposed 22 snake species belonging to seven families (Typhlopidae, Boidae, Colubridae, Natricidae, Psammophiidae, Elapidae, and Viperidae). Two families –Natricidae and Viperidae – are represented by a single species, each belonging to genera *Fowlea* and *Echis.* Four families – Typhlophidae, Boidae, Erycidae and Psammophiidae – are denoted by two species each, one species each belonging to the genera *Ramphotyphlops* and *Grypotyphlops* under the first and, *Eryx* and *Psammophis* under the third, respectively with two species. One family – Elapidae is such three species thru its place represented that to the genera *Bungarus*(two species) and Naja(one species). The family Colubridae is most specious and represented by 11 species belonging to eight genera. One species each belong to the following genera *Boiga,Coelognathus*, *Ptyas*, *Platyceps,* and *Lytorhynchus*. Two species each belong to the following genera *Oligodon,Lycodon*, and *Spalerosphis*. Though, one subspecies (*Echis carinatus sochureki*) is also representing the Viperidae family.

All snake fauna of the golden triangle is more or less evenly distributed throughout the state, except for rare to very rare species. In the Thar Desert area with similar environmental modify conditions, 80-90% of the snake species recorded in all the districts. Maximum numbers of snakes found in districts suitable habitat diversity tracts from barren dunes to canal areas interspersed with agricultural areas. As many as 21 species of the 34 species recorded from Rajasthan observed in Bikaner district, it is among the largest districts and has more or less evenly distributed habitats. Next in line was Jodhpur District (18 species), followed by Jaisalmer District (17 species and one subspecies); both these districts have arid and semi-arid regions.

Overall, it reflects that the snake diversity in this region requirements advance surveys to organize information about the present-day population, diversity, and threats to the species and their habitat preferences.

ACKNOWLEDGEMENT

This study is thankful to CSIR- New Delhi (SRF 09/098(0132)/17/EMR-I) for financial support. We also pay gratitude to Rajasthan's Forest department for the permission of this study (F3(02)/P/II/CWLW/2016/1754).

REFERENCES

- Blanford, W.T. (1901): Distribution of Vertebrate animals in India, Ceylon and Burma. *Philos. Trans. R. Soc. London*, (B) 194: 335-436.
- Das, I. (1997): Checklist of the reptiles of India with English common names. *Hamadryad, Mamallapuran*, 22 (1): 32–45.
- Husain, A. & Purohit, S. (2019): Rescue and Rehabilitation of Urban Snakes in Western Rajasthan with their Account. *International Journal of Theoretical & Applied Sciences*. 11(2): 17-28.
- Kumawat, R. & Purohit, A. (2020): Snake diversity in golden triangle of Thar Desert of Rajasthan, India. *Uttar Pradesh Journal of Zoology*. 41(24): 27-32.
- Minton, S. A. (1966): A contribution to the herpetology of West Pakistan. *Bulletin of the American Museum of Natural History.* 134(2): 27-184.
- Sharma, G. (2013): A review on the Studies on Faunal diversity, status, Threats and Conservation of Thar Desert or Great Indian Desert Ecosystem. *Biological Forum – An International Journal*. 5(2): 81-90.

- Sharma, R. C. (2003): Handbook – Indian snakes. *Zoological Survey of India, Kolkata*: i-xx + 1-292, Pls. 1-69.
- Sharma, R.C. & Rathore, N.S. (2004): Fauna of Desert national park, Rajasthan. In: Editor-Director. Conservation Area Series, No. 19: 1-135 (5 Plates) (Published by the Director, Zool. Surv. India, Kolkata).
- Sharma, R.C. (2007a): The fauna of India and the adjacent countries. Reptilia (Serpents) -111 Published by the *Director, Zool. Surv. India, Kolkata*: 410.
- Sinha, B. & Sharma, R.C. (2008): Records of *Eryx johnii* (Russell, 1801) (Ophidia: Boidae) and *Echis carinatus* (Schneider, 1801) (Ophidia: Viperidae) from the Thar desert, Rajasthan, India, with distributional notes on other snake. *J. Bombay Nat. Hist. Soc.*105 (3): 342-343.
- Whitaker, R. & Captain, A. (2004): Snakes of India. The field guide, *Draco Books, Chennai:* 481.

CHAPTER

10

Antiatherosclerotic and Hypolipidemic Efficacy of Medicinal Plants

Priyanka Riyad and Ashok Purohit
Department of Zoology, Jai Narain Vyas University, Jodhpur 342001, Rajasthan, India
Email: riyadpriyanka@gmail.com

ABSTRACT

One of the major causes of death are over the world is atherosclerosis, which is a widespread and chronic arterial disease. It occurs due to narrowing of blood vessels, which is caused by deposition of cholesterol, lipid, and other substances inside tunica intima. It also causes loss of elasticity and thickening of an arterial wall along with difficulty in blood flow. The intake of the synthetic drug causes myositis, diarrhea, gastric irritation, nausea and also disturbs the liver functioning beginning of human civilization natural products are treated very important as they are used for many purposes including medicine and they show fewer side effects. The herbal medicines show advantages like safety, acceptability, effectiveness, and affordability. Antioxidants derived from medicinal plants protect the body from oxidative damage and are used for the treatment of cardiovascular diseases as they neutralize free radicals. The present study focused on various plants which have hypolipidemic and antiatherosclerotic activity.

Keywords: Antiatherosclerotic, Hypolipidemic, Antioxidant, Myositis.

INTRODUCTION

The burden of chronic diseases among which cardiovascular disease is named is increasing rapidly worldwide. Cardiovascular or coronary heart disease is mainly related to the heart or blood vessels. Nowadays, CVD is considered a leading cause of worldwide deaths, and atherosclerosis is considered a leading cause of the cardiovascular disease (Chen *et al.*, 2020; Khan *et al.*, 2021). Chronic inflammatory disease atherosclerosis is caused by the deposition of lipoproteins in the wall of arteries. The deposition of cholesterol leads to the proliferation of various cell types such as smooth muscle cells, monocytes, and endothelial cells within the wall of arteries; that have a very bad effect on the vessel and obstructs the flow of blood, thus atherosclerosis is considered as the main cause of morbidity and mortality worldwide, especially in western society because of their lifestyle (Gonen *et al.*, 2005). Atherosclerosis may cause stroke or peripheral arterial disease by affecting the arteries which supply blood to the brain and those which supply blood to the leg.

Pathogenesis of Atherosclerosis

In the progression of atherosclerosis, a crucial and primary step is endothelial dysfunctioning (ED). An active monolayer of cells that separates the vascular wall from the circulating blood and covers the lumen of blood vessels in the vascular endothelium. The vascular homeostasis is maintained by the endothelium as it plays a key role in keeping the delicate balance between vasoconstriction and vasodilation. A metabolically active organ endothelium acts as a selective barrier for the arterial wall (Park and Park, 2015). Vascular endothelial dysfunction may cause due to hyperglycemia, hypertension, oxidative stress, increased level of low-density lipoprotein cholesterol levels, and smoking (Dagher *et al*., 2021). Chemokines and cytokines are released by injured vascular endothelial cells. Firstly the monocytes are attracted by the chemokines to the injured area through blood circulation, and then by interacting with adhesion molecules the monocytes get attached to the endothelium. Monocyte releases cytokines by penetrating the subendothelial space where it differentiates and matures into macrophages. When LDL cholesterol infiltrates the subendothelial space when the levels are high and get oxidized when retained in the intima. The accumulated oxidized LDL cholesterol forms foam cells when eaten up by macrophages. In the endothelium, various growth factors are triggered by oxidized lipids. The active production of the extracellular matrix was proliferated by vascular smooth muscle cells of the media when they transform and migrate into the intima (Katakami, 2018). Fat-rich fibrous cap around the plaque was developed due to the acceleration of smooth muscle and fibroblast growth factors, which provide stability to the plaque and reduce the blood flow in the vessels. At the boundaries of plaque macrophages and T lymphocytes are also found. Macrophages secrete the meta-proteinase, which is responsible for the lysis of the extracellular matrix. Collagen synthesis in the smooth muscle cells was prevented by TNF-α which is produced by T cells. These processes are associated with the destruction of the fibrous cap, which exposes collagen and lipids to the bloodstream which contributes to the accumulation and adhesion of platelets and blood clot formation and develops venous thrombosis. Rapture of the plaque and the extent of thrombosis are clinically associated with a wide spectrum of acute coronary syndrome and may lead to stroke, heart attacks, and death (Kopaei *et al*., 2014).

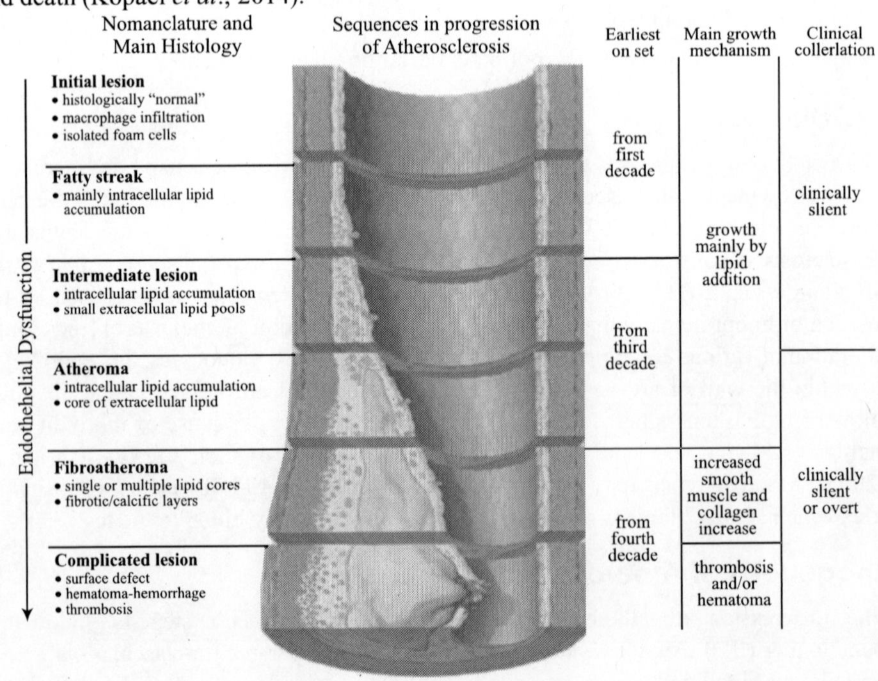

Fig. 1: *Initiation and progression of atherosclerosis* (Nelson *et al*., 2017)

The formation and evolution of atherosclerotic plaque involve several genetic, metabolic, and environmental factors. A well-known risk factor in humans is hypercholesterolemia i.e. elevated total cholesterol (TC) and low-density lipoprotein cholesterol (Purohit *et al.*, 2012). Oxidative stress also plays the important role in the progression of atherosclerosis. Oxidative stress is responsible for the formation of reactive oxygen species resulting in the oxidation of LDL particles which is promoting atherogenesis and its vascular complications (Huseini *et al.*, 2015). Other important contributors to this disease include inflammation, hypertension, smoking, and diabetes (Coto *et al.*, 2019; Subramani *et al.*, 2016).

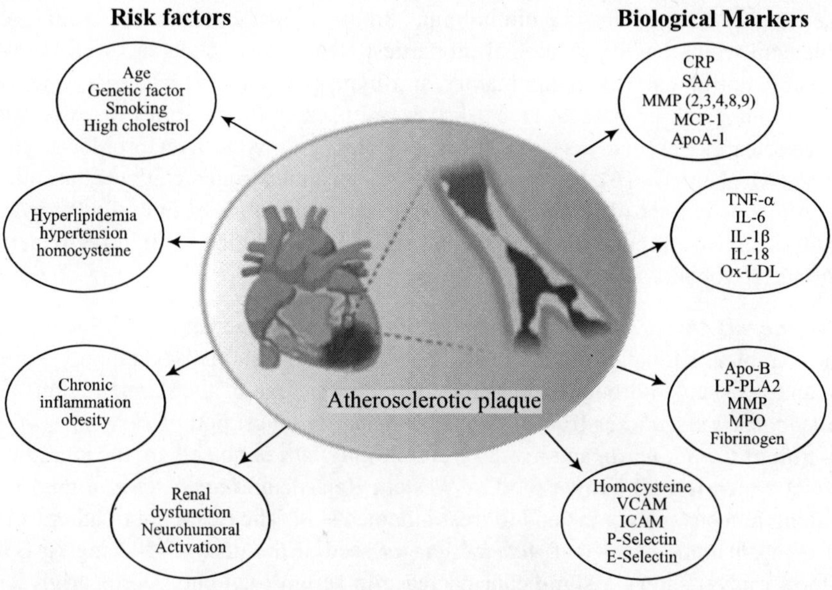

Fig. 2: *Schematic representation of different risk factors and biomarkers associated with atherosclerosis is useful in disease diagnosis and progression* (Pillai et al., 2021).

There is a range of medications available for the treatment of atherosclerosis. Reducing risk factors like taking a healthy diet, exercise and quitting smoking the process of atherosclerosis may slow down or stops. Various synthetic drugs including statins, fibrates, ezetimibe, and niacin can slow or even reverse the effect of atherosclerosis (Bergheanu *et al.,* 2017). The intake of the synthetic drug causes myositis, diarrhea, gastric irritation and also disturbs liver functioning (Thompson *et al.*, 2016). In contrast, there are many plant-based naturally generated, efficacious, and safe medicinal agents for all the population groups.

India is famed for its greater and larger biodiversity which provided a large number of medicinal plants. Indian traditional system is the oldest system which is characterized by Ayurveda, Unani, Sidha, homeopathy, and naturopathy systems, and it gives systematic information about biochemical and pharmaceutical properties of medicinal plants (Sen *et al*., 2016). The role of herbal products producing beneficial effects in disease processes and strengthening the endogenous antioxidant defenses is well established. Thus, attention is now directed to develop naturally derived safe medicinal agents for treating atherosclerosis.

Medicinal plants show various therapeutic applications to cure many diseases including metabolic disorders, neurological problems, inflammatory disease, and cardiovascular disease due to the presence of many phytochemicals such as phytosterols, polyphenols, vitamins, and flavonoids (Wei *et al.*, 2021).

Many plants have been explored for their antiatherosclerotic and hypolipidemic efficacy. Some of them are as follows:

***Azadirachta indica*:** It is a member of the Meliaceae family. It is locally known as Neem and Indian lilac. The active constituent of neem is azadirachtin. Neem also contains various types of components like nimbolinin, Nimbin, nimbanene, nimbidol, nimbiol, and ascorbic acid which having medicinal properties. Neem constituents are well known for antimalarial, antifungal, anti-inflammatory, anti-hyperglycemic, antibacterial, antioxidant, and anticarcinogenic properties. Neem leaves are used to treat various diseases such as skin ulcers, cardiovascular diseases, fever, diabetes, gingivitis, liver problems, and eye disorders (Alzohairy, 2016). Neem seed oil shows an antiandrogenic effect (Purohit *et al.*, 2008). Aqueous extract of *Azadirachta indica* leaf at the dose of 500, 1000 mg/kg/day for 28 days showed a reduction of oxidative stress and inhibition of lipid peroxidation in Wistar rats (Nwobodo, 2017).

***Acacia senegal*:** *Acacia senegal* is a member of the Mimosaceae family. It is locally known as Gum Arabica, Kumat, Kumta, and Kumatio. The tree is small deciduous, extremely hardy, and resistant to drought and main commercial crops of the desert region. Kumat seed of *Acacia senegal*, ker fruit of *Capparis decidua*, sangri pod of *Prosopis cineraria*, gunda fruit of *Cordia myxa*, and mango fruit of *Mangifera indica* all are the ingredients of Pachkutta which is well-known food of Western Rajasthan. *Acacia senegal* gum is used in soothing mucous membrane and to treat inflamed skin. The treatment of atherosclerotic rabbit with ethanolic extract of *Acacia senegal* seed at the dose of 500 mg/kg BW per day (for 45 days) shows a significant decrease in serum total cholesterol, triglycerides, LDL cholesterol, and VLDL cholesterol levels whereas HDL cholesterol increased in the experimental group. It is also beneficial for the treatment of bronchitis, malaria, diarrhea, and typhoid fever (Ram *et al.*, 2014).

***Capparis decidua*:** *Capparis decidua* belongs to the family Capparidaceae. It is commonly known as Ker. It is used for the treatment of various diseases such as toothache, arthritis, asthma, cough, inflammation, intermittent fevers, malaria, rheumatism, and swelling. The plant has antioxidant and antidiabetic activity due to the presence of amino acids, fatty acids, tocopherols, sterols, glucosinolate, and phenolic content (Zia-Ul-Haq *et al.*, 2011). Ethanolic extracts of *C.decidua* fruit, shoot, and flower shows antiatherosclerotic, hypolipidemic, and cardioprotective efficacies when treated to hyperlipidemic rabbits (Purohit and Vyas, 2005; Purohit and Vyas, 2006a; Purohit and Vyas, 2006b).

***Curcuma longa*:** *C. longa* is a medicinal plant that belongs to Zingiberaceae (ginger) family. It is usually known as Turmeric and Haldi. It is used widely as a spice in India. *C. longa* has a fat-soluble, polyphenolic pigment known as curcuminoids, which is responsible for the yellow color of turmeric. *C. longa* has antioxidant, anti-inflammatory, antiviral, and antifungal properties due to the presence of the active principle known as curcumin. Curcumin is also beneficial for the treatment of many human chronic diseases such as asthma, bronchitis, inflammatory disease, rheumatoid arthritis, coronary artery disease, atherosclerosis plaque stabilizing, diabetes mellitus, obesity, metabolic syndrome, and cancer. *Curcuma longa* also shows antifertility activities via inhibiting Leydig cells functions and arrest of spermatogenesis (Purohit and Bhagat, 2004). Treated with curcumin at the dose of 500, 1000, and 1500 mg/kg/day for 4 days then it indicating the reduced action of CD36/FAT protein and show antiatherosclerotic effect in Ldlr-/- mice (Zingg *et al.*, 2017).

***Prosopis cineraria*:** It is belonging to the family Fabaceae, usually known as Khejari, Jandi, and Kalpvariksha. It is the state tree of Rajasthan. Plants show many therapeutic properties such as anti-inflammatory, antidiabetic, antiasthmatic, anticancer due to the presence of many active compounds including alkaloids, polyphenols, phenolic, compounds saponins. Bark extract (500 mg/kg body weight/day) of *P.cineraria* shows a significant reduction in lipid profile and also reduces the plaque area and thickness of arterial intima when treated to hyperlipidemic rabbits for 45 days (Purohit and Ram, 2012; Ram *et al.*, 2020).

***Cuminum cyminum* L.:** It belongs to the Apiaceae family, also known as cumin. Like many other spices, cumin seeds are also used as a traditional ingredient in the Indian kitchen. Cuminaldehyde is considered a bioactive compound of cumin. Cumin has been used for curing many diseases like digestive disorders, chronic diarrhea, diabetes, cancer, cardiovascular disease, and dyspepsia. Ethanolic extract of cumin seed also shows hypolipidemic activity (Chouhan and Purohit, 2018; Srinivasan, 2018).

***Triticum aestivum*:** It is a Poaceae family member, popularly known as wheat. Wheat is the largest consuming cereal grain all over the world because it contains proteins, vitamins, and minerals. Wheat is a good source of phytochemicals such as lignin, phenolic acids, phytosterols, carotenoids, etc. It can decrease the risk of cardiovascular disease, cancer, diabetes, and manage body health. Administration of wheat wholegrain extract and wheat germ oil in hyperlipidemic rabbits showed a significant reduction of serum total cholesterol, LDL-C, VLDL-C, triglycerides, and atherosclerotic plaque area (Chadha *et al.*, 2015; Chadha and Purohit, 2018).

Acacia nilotica: *Acacia nilotica* is belonging to the family Mimosaceae. It is also known as Babul and Bambul. In comparison to diabetic rabbits with that of treated rabbits, reduction in total cholesterol, blood glucose, triglycerides, the plasma, and low-density lipids (LDL's) were seen when they were given a dose of 400 mg/kg of body weight. Properties of this plant-like antidiabetic and antiatherosclerotic potential were shown by this study (Batra *et al.*, 2011).

Momordica charantia: *Momordica charantia* belongs to the Cucurbitaceae family. It is locally named Karela and Bitter guard. This plant is known as a medicinal plant as it is used to treat many diseases like cancer, diabetes mellitus, infections, and fever in Ayurveda. A dose of 1& 1.5 ml/kg by body weight of fresh juice is taken for 10 days then it indicates the hypolipidemic, hypoglycemic, and antioxidant activity in Norwegian rats (Sharmin *et al.*, 2017).

***Trigonella foenum-graecum*:** It is an annual herb that belongs to the family Fabaceae. It is commonly known as Dana Methi or Fenugreek. Fenugreek seeds are the main ingredient of Indian dishes. The seed contains proteins, dietary fibers, vitamins, and other dietary minerals. Active constituents of fenugreek are trigonelline and saponins. Fenugreek has hypocholesterolemic, antidiabetic, antioxidant, hepatoprotective, anticancer effects, and digestive stimulant action (Fatima *et al.*, 2018). It demonstrated that the aqueous solution of seed powder at the dose of 500mg/kg/day on Albino rabbits possesses potent antihyperglycemic and hypolipidemic activity (Sharma and Choudhary, 2017).

***Cinnamomum tamala*:** It is a member of Lauraceae family. It is locally known as Tejpatta or Indian bay leaf. Active constituents of *C. tamala* are alkaloids, phenolic compounds such as linalool, beta caryoohyllene. The aqueous and ethanolic extract of *C. tamala* leaf at the dose of 250 mg/kg/day has been reported to possess significant hypoglycemic and hypolipidemic effects in spargue dawlery rats (Singh *et al.*, 2018).

***Zingiber officinale*:** It is belonging to the Zingiberaceae family. *Zingiber officinale* is locally known as Adrak, Sounth, and Ginger. Ginger plays an important role in traditional Indian Ayurvedic medicine. Ginger contains volatile oils, zingerone, and gingerols as an active principle which are known for anti-inflammatory, cardioprotective, hypolipidemic, hypoglycemic, antioxidant, and antimicrobial activity of ginger. Ginger is used to curing several diseases such as motion sickness, osteoarthritis, nausea, epilepsy, inflammation, food poisoning, cancer, cough, and cold. Hydroethanolic extract of *Zingiber officinale* bulb at the dose of 250 mg/kg/day for 5 weeks showed hypolipidemic effect due to high expression of PPARα, PPAR-γ and glut-2 in male Wistar rats (Heras *et al.*, 2016).

Withania somnifera: It is a member of the Solanaceae family. It is locally known as Ashwagandha. It is an Ayurvedic medicinal plant that is used for several ailments. Active phytoconstituents of *W. somnifera* are withanolides, withaferin-A, and other phytochemicals which have a positive effect on hyperlipidemia. Hypoglycaemic and hypolipidemic effects of alcoholic extracts of root and leaves were investigated at the doses of 50 and 100 mg/kg body weight in albino rabbits. After treatment with the extract, a significant decrease was observed in the level of blood glucose, serum cholesterol, LDL, and VLDL (Laylani and Sateh *et al.*, 2018).

***Ocimum sanctum*:** *O. sanctum* is an aromatic shrub belonging to the family Lamiaceae. It is commonly known as Tulsi and holy basil. Tulsi has medicinal and spiritual properties. It is an important Ayurvedic medicinal shrub. In Ayurveda, it's called "The unequaled One," "Mother drugs of Nature" and "The Queen of Herbs". Active compounds of *O. sanctum* are ocimumosides and ocimarin. Tulsi is used for the treatment of various diseases, including bronchitis, malaria, diarrhea and dysentery, anxiety, cough, asthma, inflammation, skin diseases, arthritis, cancer, and cardiovascular disorders. Several clinical studies were carried out to confirm the therapeutic potential of *O. sanctum* as antimicrobial, mosquito repellent, anti-oxidant, anti-inflammatory, hepato-protective, neuro-protective, cardio-protective, anti-diabetic, anti-hypercholesterolemia, anti-carcinogenic, analgesic, anti-allergic, memory enhancement, and anti-asthmatic (Cohen, 2014). The ethanolic and aqueous extract of *O. sanctum* leaves has been shown to cause significant reduction of blood glucose, total cholesterol and also showed anti-hyperlipidemic activity in streptozotocin-induced DM rats (Suanarunsawat *et al.*, 2016).

Coriandrum sativum **(L.):** Belongs to the family Apiaceae. Popularly known as coriander. It is the main ingredient of Indian recipes. Apigenin, quercetin, kaempferol, vanillic acid, and ferulic acid are the main bioactive compounds of coriander. Antimicrobial, antifertility, hyperglycemic, and cardioprotective activities are shown by coriander. Coriander seed also shows antiatherosclerotic and antioxidant activities via reducing serum total cholesterol, phospholipid, LDL-cholesterol, triglycerides, catalase, and glutathione activity and increase the hepatic lipid peroxidation (Joshi *et al.*, 2012).

Table: Summarized list of plants that have been reported for their Hypolipidemic and Antiatherosclerotic potential

S. No.	Name of the plant	Family	Common name	Plant parts used	Bioactive constituents	Antiatherosclerotic and hypolipidemic effects
1.	*Azadirachta indica*	Meliaceae	Neem and Indian lilac	Leaf	Nimbolinin, nimbin, nimbanene, nimbidol, nimbiol and ascorbic acid	Reduce oxidative stress and inhibit lipid peroxidation.
2.	*Acacia senegal*	Mimosaceae	Gum Arabica, Kumat	Seed	Flavonoid, tannin, phenolic acid.	Significantly reduce total serum cholesterol, LDL-C, VLDL-C, and triglyceride levels.
3.	*Capparis decidua*	Capparidaceae	Ker	Fruit, shoot, flower	Amino acids, fatty acids, tocopherols, sterols, glucosinolate, and phenolic compounds	Significantly reduce total serum cholesterol, LDL-C, VLDL-C, and triglyceride levels.
4.	*Curcuma longa*	Zingiberaceae	Turmeric and Haldi	Bulb	Curcumin	Reduce lipid profile Reduce the action of CD36/FAT protein
5.	*Prosopis cineraria*	Fabaceae	Khejari, Jandi, and Kalpvariksha	Bark	Alkaloids, polyphenols, phenolic acids, saponins	Significantly reduce lipid profile and atherosclerotic plaque area
6.	*Cuminum cyminum*	Apiaceae	Cumin	Seed	Cuminaldehyde	Decrease serum total cholesterol, LDL-C, VLDL-C, and triglyceride level
7.	*Triticum aestivum*	Poaceae	Wheat	Whole grain and germ oil	Lignin, phenolic acids, phytosterols, carotenoids	Reduce cholesterol and atherosclerotic plaque area
8.	*Acacia nilotica*	Mimosaceae	Babul and kikar	Pods	Lupenone, lupeol, niloticane	Significantly reduce lipid profile
9.	*Momordica charantia*	Cucurbitaceae	Karela and Bitter guard	Fruit	Glycosides, saponins, alkaloid, triterpenes, and proteins	Decrease serum total cholesterol, LDL-C, VLDL-C, and triglyceride level

10.	*Trigonella foenum-graecum*	Fabaceae	Dana Methi, Fenugreek	Seed	Trigonelline and saponins	Reduce lipid profile
11.	*Cinnamomum tamala*	Lauraceae	Tejpatta or Indian bay leaf	Leaf	Linalool, beta caryoohyllene	Decrease serum total cholesterol, LDL-C, VLDL-C, and triglyceride level
12	*Zingiber officinale*	Zingeberaceae	Adrak, Sounth, and Ginger	Bulb	Volatile oils, zingerone, and gingerols	Increase the expression of PPARα, PPAR-γ and glut-2
13.	*Withania somnifera*	Solanaceae	Ashwagandha	Root, leaf	Withanolides, withanoferin-A	Reduce cholesterol, LDL, and VLDL
14.	*Ocimum sanctum*	Lamiacea	Tulsi and holy basil	Leaf	Ocimumosides and ocimarin	Significantly reduce lipid profile
15.	*Coriandrum sativum*	Apiaceae	Coriander	Seed	Quercetin, kaempferol, apigenin, vanillic, and ferulic acids	Increase hepatic lipid peroxidation and reduce glutathione and catalase activity

CONCLUSION

Herbal products are well known for their pharmaceutical and nutraceutical properties since ancient times. The medicinal uses of plants are documented in ancient traditional systems. Previous study data indicate that the use of plant-based medicine for the treatment of various ailments such as diabetes, myopathy, cancer, inflammation, and metabolic diseases, having beneficial and adverse effects as compared to allopathic drugs. Thus, the use of herbal medicines is continued in modern society for the prevention and treatment of cardiovascular diseases and currently available drugs are also mainly derived from the plant. A large number of plants are explored for their antiatherosclerotic activity by reducing serum total cholesterol, low-density lipoprotein cholesterol, and triglycerides levels. A further comprehensive study of the mechanisms of influence of natural products on key links in the chain of development of atherosclerosis is an important and necessary step on the way to the development of new safe and effective preparations based on medicinal plants.

ACKNOWLEDGMENT

We are thankful to the Head and Department of Zoology, Jai Narain Vyas University, Jodhpur, Rajasthan and UGC. Heartfelt thanks to Professor Ashok Purohit and Professor Kalpana Purohit for inspiring us to write this paper. Special thanks to Dr. H. Ram Barad and Dr. R.P. Saran.

REFERENCES

- Alzohairy, M.A. (2016): Therapeutics role of *Azadirachta indica* (Neem) and their active constituents in diseases prevention and treatment. *Evidence-Based Complementary and Alternative Medicine,* 7382506:1-11.
- Batra, S., Praksah, N.B. and Batra, N. (2011): Medicinal plants of Rajasthan (India) with antidiabetic activity. *International Research Journal of Pharmacy,* 2(3):1-7.

- Bergheanu, S.C., Bodde, M.C., and Jukema, J.W. (2017): Pathophysiology and treatment of atherosclerosis: Current view and future perspective on lipoprotein modification treatment. *Netherlands Heart Journal*, 25(4):231-242.
- Chadha, R., Ram, H., and Purohit, A. (2015): Hypolipidemic effects of wheat germ oil in cholesterol-fed rabbits. *Medicinal Drug Research*, 3(2):15-20.
- Chadha, R. and Purohit, A. (2018): Lipid-lowering antioxidant effect of wheat wholegrain extract in hypercholesterolemic rabbits. *European Journal of Biomedical and Pharmaceutical Sciences*, 5(1):541-546.
- Chen, J., Zhang, X., Millican, R., Creutzmann, J.E., Martin, S., and Jun, H.W. (2020): High density lipoprotein mimicking nanoparticles for atherosclerosis. *Nano Convergence*, 7(1):6.
- Chouhan, H. and Purohit, A. (2018): Protective effect of cumin (*Cuminum cyminum* L.) seed extract on the cardiovascular system, toxicity, and hematology on hyperlipidemic rabbits: An experimental study. *Asian Journal of Pharmaceutical and Clinical Research*, 11(10):353-356.
- Cohen, M.M. (2014): Tulsi-*Ocimum sanctum*: A herb for all reasons. *Journal of Ayurveda and Integrative Medicine*, *5*(4):251-9.
- Coto E., Fernandez, N., Garcia, J.J., Diez, M.J., Sahagun, A.M. and Sierra, M. (2019): Assessment of the antioxidant/hypolipidemic relationship of *Sideritis hyssopifolia* an experimental animal model. *Molecules* (*Basel, Switzerland*), 24(11), 2049.
- Dagher, O., Mury, P., Thorin-Trescases, N., Noly, P.E., Thorin, E., and Carrier, M. (2021): Therapeutic potential of quercetin to alleviate endothelial dysfunction in age-related cardiovascular diseases. *Front. Cardiovasc. Med.*, 8:658400.
- Fatima, T., Maqbool, K. and Hussain, S.Z. (2018): Potential health benefits of fenugreek. *Journal of Medicinal Plants Studies*, 6(2):166-169.
- Gonen, A., Harats, D., Rabinkov, A., Miron, T., Mirelman, D., Wilchek, M., Weiner, L., Ulman, E., Levkovitz, H., Ben-Shushan, D., and Shaish, A. (2005): The antiatherogenic effect of allicin: Possible mode of action. *Pathobiology*., 72:325-334.
- Heras, N.D., Munoz, M.V., Fernandez, B.M., Ballesteros, S., Farre, A.L., Ruiz-Roso, B. and Lahiri, V. (2016): Molecular factor involved in the hypolipidemic and insulin sensitizing effects of a ginger (*Z. officinale*) extract in rats fed a high fat diet. *Applied Physiology, Nutrition, and Metabolism,* 42(2):209-215.
- Huseini, H.F., Anvari, M.S., Khoob, Y.T., Rabbani, S., Sharifi, F., Arzaghi, S.M., and Fakhrzadeh, H. (2015): Anti-hyperlipidemic and anti-atherosclerotic effects of *Pinus eldarica* Medw. nut in hypercholesterolemic rabbits. *DARU Journal of Pharmaceutical Sciences*., 23:32.
- Joshi, S.C., Sharma, N., and Sharma, P. (2012): Antioxidant and lipid-lowering effects of *Coriandrum sativum* in cholesterol-fed rabbits. *International Journal of Pharmacy and Pharmaceutical Sciences*, 4(3):231-234.
- Katakami, N. (2018): Mechanism of development of atherosclerosis and cardiovascular disease in diabetes mellitus. *Journal of Atherosclerosis and Thrombosis*, 25(1): 27-39.
- Khan, J., Deb, P.K., Priya, S., Medina, K.D., Devi, R., Walode, S.G., and Rudrapal, M. (2021): Dietary flavonoids: Cardioprotective potential with antioxidant effects

and their pharmacokinetic, toxicological and therapeutic concerns. *Molecules (Basel, Switzerland)*, 26(13):4021.

- Kopaei, M.R., Setorki, M., Doudi, M., Baradaran, A., and Nasri, H. (2014): Atherosclerosis: Process, indicators, risk factors, and new hopes. *International Journal of Preventive Medicine*, 5(8):927-46.
- Laylani, L.A., and Saleh, A.H. (2018): Alcoholic extract effect of *Withania somnifera* roots on cholesterol diet-induced hypercholesterolemia in male rabbits. *Iraqi Journal of Science,* 58(1):267-270.
- Nelson, J.R., Wani, O., May, H.T., and Budoff, M. (2017): Potential benefits of eicosapentaenoic acid on atherosclerotic plaques. *Vascular Pharmacology*, 91(2017): 1-9.
- Nwobodo, E.I. (2017): Evalution of antilipid peroxidation and hypolipidemic potential of *Azadirachta indica* leaf aqueous extract in paracetamol-induced hepatotoxicity in Wistar rats. *International Journal of Information Research and Review,* 4(2):3615-3619.
- Park, K.H. and Park, W.J. (2015): Endothelial dysfunction: Clinical implications in cardiovascular disease and therapeutic approaches. *Journal of Korean Medical Science*, 30(9):1213-25.
- Pillai, S.C., Borah, A., Jacob, E.M., and Kumar, D.S. (2021): Nanotechnological approach to delivering nutraceuticals as promising drug candidates for the treatment of atherosclerosis. *Drug Delivery*, 28(1):550-568.
- Purohit, A. and Bhagat, M. (2004): Contraceptive effect of *Curcuma longa* (L.) in male albino rats. *Asian Journal of Andrology*, 6(1):71-74.
- Purohit, A., Joshi K., Kotru, B., and Kotru, S. (2012): Hypolipidemic effects of bees wax on serum lipid parameters in induced hyperlipidemic male albino rabbits. *Journal of Cell and Tissue Research*, 12(3):3311-3316.
- Purohit, A., Joshi V.B., and Vyas, K.B. (2008): Effect of various chromatographic fractions of neem seed oil on sperm dynamics and testicular cell population dynamics on male albino rats. *Pharmaceutical Biology*, 46(3):660-664.
- Purohit, A. and Ram, H. (2012): Hypolipidemic and antiatherosclerotic effects of *Prosopis cineraria* bark extract in experimentally induced hyperlipidemic rabbits. *Asian Journal of Pharmaceutical and Clinical Research*, 5(3):106-109.
- Purohit, A. and Vyas, K.B. (2006a): Antiatherosclerotic effects of *Capparis decidua* fruit extract in cholesterol-fed rabbits. *Pharmaceutical Biology*, 44(3):172-177.
- Purohit, A. and Vyas, K.B. (2006b): Antiatherosclerotic efficacy of *Capparis decidua* flower extract in cholesterol-fed rabbits. *Journal of Cell and Tissue Research*, 6(1):533-536.
- Purohit, A. and Vyas, K.B. (2005): Hypolipidemic efficacy of *Capparis decidua* fruit and shoot extracts in cholesterol-fed rabbits. *Indian Journal of Experimental Biology*, 43:863-866.
- Ram, H., Jatwa, R., and Purohit, A. (2014): Antiatherosclerotic and cardioprotective potential of *Acacia senegal* seeds in diet-induced atherosclerosis in rabbits. *Biochemistry Research International,* 436848:1-6.

- Ram, H., Jaipal, N., Charan, J., Kashyap, P., Kumar, S., Tripathi, R., Singh, B.P., Siddaiah, C.N., Hashem, A., Tabassum, B., and Abd-Allah, E.F. (2020): Phytoconstituents of an ethanolic pod extract of *Prosopis cineraria* triggers the inhibition of HMG-CoA reductase and the regression of atherosclerotic plaque in hypercholesterolemic rabbits. *Lipids in Health and Disease*, 19:6.
- Sen, S., Chakraborty, R., and De, B. (2016): Indian traditional systems, herbal medicine, and diabetes. Diabetes mellitus in 21st century. *Springer, Singapore.*, 978-981.
- Sharma, M.S., and Choudhary, P.R. (2017): Effect of fenugreek seeds powder (*Trigonella foenum-graecum*) on experimental induced hyperlipidemia in rabbits. *Journal of Dietary Supplements.*, 14(1):1-8.
- Sharmin, Z.R., Akter, M.T., Rahman, R., Hoque, M., Khan, I. and Mosaddek, A.S. (2017): Comparative study of hypolipidemic effect of *Momordica charantia* (Karela) with atorvastatin in fat-fed rats. *Journal of Human Health Research,* 1(1):11-15.
- Singh, V., Singh, S.P., Singh, M. and Kumar, A. (2018): Evalution of antioxidant effects of the phytoconstituents of *Cinnamomum tamala* in rats. *Indian Journal of Pharmaceutical Sciences*, 80(1):161-72.
- Srinivasan, K. (2018): Cumin (*Cuminum cyminum)* and black cumin (*Nigella sativa*) seeds: traditional uses, chemical constituents, and nutraceutical effects. *Food Quality and Safety*, 2(1):1-16.
- Suanarunsawat, T., Anantasomboon, G., and Piewbang, C. (2016): Anti-diabetic and anti-oxidative activity of fixed oil extracted from *Ocimum sanctum* L. leaves in diabetic rats. *Experimental and Therapeutic Medicine*, 11(3):832-840.
- Subramani, C., Rajakkannu, A., Rathinam, A., Gaidhani, S., Raju, I. and Kartar Singh, D.V. (2016): Anti-atherosclerotic activity of root bark of *Premna integrifolia* Linn. in high fat diet induced atherosclerosis model rats. *Journal of Pharmaceutical Analysis*, 7(2):123-128.
- Thompson, P.D., Panza, G., Zaleski, A. and Taylor, B. (2016): Statin associated side effects. *Journal of American College of Cardiology,* 67(20):2395-2410.
- Wei, T., Liu, J., Zhang, D., Wang, X., Li, G., Ma, R., Chen, G., Lin, X., and Guo, X. (2021): The relationship between nutrition and atherosclerosis. *Front. Bioeng. Biotechnol.*, 9:635504.
- Zia-Ul-Haq, M., Cavar, S., Qayum, M., Imran, I., and de Feo, v. (2011): Compositional studies: antioxidant and antidiabetic activities of *Capparis decidua* (Forsk.) Edgew. *International Journal of Molecular Sciences*, 12(12):8846-61.
- Zingg, J.M., Hasan, S., Nakagawa, K., Canepa, E., Ricciarelli, R., Villacorta, L., Azzi, A. and Meydani, M. (2017): Modulation of CAMP levels by high-fat diet and curcumin and regulatory effects on CD36/FAT scavenger receptor/fatty acids transporter gene expression. *Biofactors,* 43(1): 42-53.

CHAPTER 11

Role of Traditional Medicinal Plants in Fertility Regulation of Male and Female—A Safe Approach and Current Need

Nisha Kanwar, R.P. Saran, Ashok Purohit
Department of Zoology, Jai Narain Vyas University, Jodhpur, 342001, Rajasthan, India
Email: nishakanwar25@gmail.com

ABSTRACT

Aim of this review is to study different researches done on the fertility regulation properties of different medicinal plants. Nowadays birth control and solving fertility related problems i.e. fertility regulation in human beings is becoming a major warning issue. Present review is on scientific researches done on medicinal plants used for fertility regulation. This review presents knowledge of such medicinal plants with its family, common name, botanical name, parts used, dose used, extract used and solvent used for research. This review revels that there are ample number of plants present with the fertility regulation activity but few of them are studied though researches for contraceptive activity i.e. anti fertility activity are in ample numbers. Studies on treatment of fertility related issues are very few which need to be evaluated. Detailed study of such plants and documentation is needed for the formulation of herbal product which may be used by humans.

Keywords: Ayurveda, secondary metabolites, medicinal plants, plant extracts, fertility regulation, antifertility, overpopulation, traditional medicines.

INTRODUCTION

Over past 100 years rapid growth in the world's population has been registered. This growth of population is due to difference in birth rate and death rate. After the medicinal or industrial revolution which improved the nutritional value, sanitation and medicinal facilities. Due to all these improvements more infants are surviving in their early year of life as compared to the early decades. This difference between high birth rate and low death rate has counted to increased population. It is one of the reasons of over population. Over population is the condition where more people that can thrive on earth are living. Greater number of people with discomfort, unhappiness, health issues and will leave the world unfit place for future generation to live. In many countries reproductive health problems are becoming prominent. In 2002 World Health Organisation (WHO) illuminated different problems which are rising at warning rate such as birth asphyxia, maternal illness and difficulties after birth of child (Adhikari *et al.*, 2018). Overpopulation is a global problem which is costing economic, social and natural recourses loss.

Tradition medicines are used by man since time immemorial. Most of the tradition medicines are based on plants. The knowledge about different medicinal plants is present in Ayurveda, unani and siddha. Due to problems and side effects associated with synthetic drugs and emergence of multi drug resistant pathogens. The world is witnessing interest in the study of traditional medicine plants (Chellaiah *et al.*, 2006).

In India plants have always played important role in health care sector due to cost effective and easy availability. Medicinal plants and their use in treatment of various diseases is indispensable part of tradition.

In Rigveda use of medicinal plants is the earliest record present (Kunwar *et al.*, 2006). This written literature is between 4500BC and 1600BC. In this literature 67 plants are mentioned and it is the oldest one. In Rig-Veda plants are mentioned as "God for Gods" (Malla *et al.*, 1984).

The ancient Indian literature is Rigveda in which several herbs are listed which have medicinal (healing) properties but more detailed account is found in the Atharvaveda. Ayurveda (Ayus means life and Veda means Knowledge) which is Indian traditional health care system is known to be originated from Athervaveda. Athervaveda has more detailed account of medicinal plants than Rigveda. System of medicines in Ayurveda is based on two scripts or thesis one is Charak Samhita and other is Shusruta Samhita (Parchure, 1983, Sharma, 1993).

70-80 percent of world population relies on medicines which are plant based or are plant products mentioned in a survey conducted by World Health Organisation (WHO) (Chan, 2003).

The demand of these tradition medicines has increased in last few decades. The idea of using higher plants for human use is not new. Use of higher plants for development of safe, orally and effective fertility regulating agents is an older practice. Plants are used from centuries in different forms to limit the population. There are numerous plants which have properties that regulate fertility. Efforts are in process to develop anti fertility and aphrodisiac products from plants. As the plant based medicines from crude extracts are especially beneficial to the poor population of the country as the cost of such medicines are affordable by the poor population. All over the world extensive researches are being carried out to evaluate the anti fertility and aphrodisiac activities of different plants. In both male and female anti fertility agents are the agents which prevent fertility by interfering in reproductive mechanisms. A drug which possess 100 percent efficacy with action which is reversible, is free of side effects and whose use is easy (John, 1981). Due to low income in these traditional medicines, lack of return documents and scarcity of scientific bases has led this traditional knowledge to disappear with time.

Agents who are capable of hampering fertilization or ovulation and also terminate pregnancy are known as antifertility agents (Shaik *et al.*, 2014).

Mechanism of action of antifertility plant products or compounds: There are various mechanisms of action present for different antifertility agents particularly for suppressing fertility major action is on sex hormones, symptoms during menopause, menstrual cycle regulation etc. (Williamson *et al.*, 1996). Plants which posse's estrogenic properties directly affect pituitary actions by attuning secretions of FSH (Follicle Stimulating Hormone) and LH (Luteinizing Hormone) and ovulation blockage (Brinker, 1997). Plants have abortificant effect as they interrupt in formation of ovum and endometrium (Gaek *et al.*, 1978, Prakesh *et al.*, 1985). In females the action site of antifertility agents includes anterior pituitary, oviduct, ovary, hypothalamus, vagina and uterus. Main site for antifertility agents in mammals is uterus (Williamson *et al.*, 1996).

Antifertility effect on male by plant extract includes effect on male reproductive system which may have action on post testicular antifertility effect, sperm immobilization effect, spermicidal effect, antiandrogenic effect and antispermatogenic effect (Shaik *et al.*, 2017).

There are many plants which are known for their spermicidal, antiimplantation, ecobolic, oestrogenic and abortification properties which are used by females as antifertility drug. (Marles *et al.*, 1994).

Traditional uses of plants as contraceptive is well known but plants are also used to cure infertility related problems. Couples who experience infertility involve 60 percent male factor which is responsible for the problem and out of left 40 percent 20 percent is due to both male and female factor and rest 20 percent is also primarily male factor which leads to infertility (Dhandpat *et al.*, 1985). So there is need for the search of such herbal products which can resolve the infertility related problems.

There are various reproductive dysfunctions in males which leads to infertility are premature ejaculation, retarded or inhibition of ejaculation, retrograded inability to achieve sexual intercourse, erectile dysfunction etc. (Yakubu *et al.*, 2007). Reduced male fertility is also due to the abnormality in sperm parameters such as decreased sperm count (oligozoospermia), decreased sperm morality (asthenozoospermia) and morphology (terratozoospermia) (Dohle *et al.*, 2005).

The available management options present in the market are associated with some side effects such as hormonal imbalance, hypertension, increased risk of cancer and weight gain (Vaidya *et al.*, 2006), expensive and not easily available for the rural and poor population. The solution for above problem is in searching medicinal plants with fertility regulating properties which can be later formulated in form of drug which is easily available with reduced cost and lesser or no side effects (Yakubu *et al.*, 2007). Traditional practises in world's most parts, India and Ethiopia used are plant based medicines (Marles *et al.*, 1994). Important constituent of reproductive health is fertility regulation which comprises of managing infertility and contraception (Allag and Rangari, 2002). To discover non steroidal contraceptive compound which is orally active and is plant based WHO has set up a task force (Montaserti *et al.*, 2007). Efficient birth control methods are the prime need for countries like India where population is increasing alarmingly. The expected increase in population of India by 2050 is estimated to be 9.2 billion (United Nations World Population Prospect, 2007).

The present review is based on such plants which have fertility regulation in both male and female which is based on numerous studies present in literature available around the world. The aim of this review was to study the available data on plants which possess the properties of anti fertility and in fertility treatment i.e. fertility regulation.

Fertility Regulation in Males

***Martynia annua* L.** (Devil's Claw) belongs to family Martyniaccae. Male rats were studied for the reproductive activity of *Martynia annua* L. root extract in 50% ethanol rats were administered with root extract at dose level of 50 mg/kg body weight, po, 100 mg/kg body weight, po, and 200 mg/kg body weight, po, respectively, for a period of 60 days. Decrease in weight of epididymis, testid, ventral prostate and seminal vesicle was observed in treated rats.

Reduction of epididymal sperm count, testicular sperm count and number of fertile males, motility, number of pups and ratio between inseminated and delivered females was observed. Male reproduction was effected without altering metabolism of body by the 50% ethanol extract of *M. annua* root (Mali *et al.*, 2002).

Maytenus emarginata *(*Kankhera*)* belongs to family Celastraceae. It is used for the treating various disorders such as Heart ailments, digestive ailments, inflammation, nerve afflictions, nervous system, Rashes and as anti stress and tonic. Albino rats were administered orally with leaf extract of *Maytenus emarginata* at dose level of 50,100,200 mg/kg/ body wt/day per oral dose for 60 days. In comparison to control decrease in sperm count and mortality of sperms was observed in the treated rats. 80% reduction in pregnancy rate was observed in females after mating with treated male with leaf extract dose of 50,100,200 mg/kg of *Maytenus emarginata*. Testicular weight and seminal vesicle also decreased (Sharma and Mali, 2016).

Opuntia elatior (Prickly Pear) belongs to family Cactaceae. Male rats were administered orally with methanol fruit extract at dose of 300, 900 mg/kg body weight for 60 days. 75-80 % reduction in epididymal sperm count and mortality was observed in treated rats for 60 days but serum testosterone level did not decline. Hydroxysteroid dehydrogenase activity study suggested that there was no effect on testicular steroidogenesis. Treated rats when mated with normal cycling female rats there fertility rate suppressed. Suppression of fertility was dose dependent 100% in highest dose. Recovery of testicular HSDH activity, serum testosterone, epididymal sperm count and fertility was observed when treatment was withdrawal (Ramya *et al.*, 2015).

***Nigella sativa* Linn.** (blackseed or black cumin) belongs to family Ranunculaceae. Seeds are used as herbal medicine all over the world for the treatment and prevention of diseases and conditions like asthma, diarrhoea and dyslipidaemia. Adult male albino rats were treated at a dose of 300.mg/kg body weight for 60 days. Seeds treatment increased the weight of reproductive organs in comparison to control rats ($P<0.01$). Sperm count and mortality in cauda epidydimides and testicular ducts were significantly increased ($P<0.01$). Seminal vesicle and ventricular prostate also increased the secretary activity. Female rats impregnated by males receiving treatment increased in number. The aqueous extracts of *Nigella sativa* have increased spermatogenesis of male albino rats (Mohammad *et al.*, 2009).

Terminalia chebula (Myrobalan, Hardad) belongs to family Combretaceae. It *in vitro* at 30 mg/ml has hyaluronidase (HAase) inhibitory activity of rat caudal epididymal spermatozoa (~86% inhibition) and human spermatozoa (~93% inhibition). Male albino rats administered at 50 and 100 mg/kg/day orally for 60 days posses antispermatogenic activity and contraceptive efficacy. Epididymal spermatozoa showed decrease in count and mortality and morphogenic abnorlmalities increased. Male rates treated with fruit extract of *T. chebula* at 100 mg/kg dose showed reduction in fertility (-100%) (Srivastav *et al.*, 2010).

Albizia lebbeck (Indian Siris) belongs to family Fabaceae (Leguminosae). Its bark methanolic extract was administered orally to male rats for 60 days. Decrease in weight of reproductive organs (seminal vesicle, epididymides, ventral prostate and testis was observed without any decrease in body weight as compared with control. Sperm density and sperm mortality were reduced which resulted in male fertility reduction by 100%. Spermatogenesis arrest was observed in male rats without any side effects when treated with bark extract of *A. lebbeck* (Gupta *et al.*, 2006).

Aegle marmelos (Golden Apple, Bael) belongs to family Rutaceae. Male albino rats were treated with *Aegle marmelos* leaf extract at doses 100, 200 and 300 mg^{-1} kg^{-1} day^{-1} for each rat for 60 days. After 120 days recovery was also tested. Treated animals showed reduction in sperm density and mortality derived from cauda epididymis. Reduction in fertility observed was 100% in males treated with *A. marmelos*. Decreased serum testosterone levels were observed in all groups. After 120 days of withdrawal of extract restoration of all altered parameters was observed (Chauhan *et al*., 2007).

Dactyloctenium aegyptium (Crowfoot Grass, Beach Wiregrass) belongs to family Poaceae. Male albino rats were used to study effect of ethanol extract *D. aegyptium* on the fertility. Ethanolic extract of D. aegyptium at a dose of 200,400, and 600 mg/kg body weight was administered for a period of 30 days. Decrease in weight of accessory sex organs, sperm count and testis, increase in sperm mortality was observed in the treated animals. Hormone levels decreased in serum. Number of implantations in female rats also decreased all results obtained indicated antifertility activity of *D. aegyptium* (Naik *et al*., 2016).

Butea monosperma (flam of forest or palash) belongs to family Fabaceae. Male albino rats were administered with methanolic extract of *Butea monosperma* leaf at dose level of 200 and 400 mg/kg body weight for 21 days. Extract reduced sperm count, mortality and also decreased reproductive organs weight in male rats (Reddy and Rajendra., 2018).

Coccinia indica (Ivy Gourd) belongs to family cucurbitaceae. Its leaf ethanolic extract was tested on rats at dose level of 200 and 500 mg/kg body weight per day for 56 days. Rats administered with extract showed decreased level of testosterone and steroidogenic markers in testis. Adverse effect was also observed in cauda epididymidis sperms (Verma and Singh., 2017).

Withania somnifera (Ashwagandha) belongs to family Solanaceae. Decrease in reproductive organs on treatment with crude plants hydroalcoholic extract was observed. Sperm density and motility was also reduced. Study suggests male infertility caused by this plant. Study also revels testosterone and LH (Luteinizing Hormone) decreased in hormonal assay i.e. the extract is working as antispermatogenic (Mali, 2013).

Cuminum cyminum (Cumin) belongs to family Apiaceae. Fruit extracts of *S. emarginatus, Allium cepa* and *T. belerica* and seed extract of *Cuminum cyminum* were tested on male albino rats. Extract was fed orally for 60 days at dose level of 50mg/day/ rat. Treatment shows decrease in reproductive organs weight without affecting the weight of body. Sperm count and motility of cauda epidymis and testis decreased (Venketesh *et al*., 2002).

Mentha arvensis (Pudina) belongs to family Lamiaceae. Petroleum ether extract of *Mentha arvensis* leaves was tested on male albino rat. Extract was orally given for 20, 40, and 60 days at dose level of 10 and 20 mg/ rat/ day. When normal females mated with treated males number of offsprings reduced in a dose and duration dependent manner. Study suggests that leaves specially the petroleum ether extract has property of reversible antifetility without any toxicity (Sharma and Jacob, 2001).

Ocimum gratissimum (Tulasi) belongs to family Lamiaceae. Due to high medicinal value of this plant it is widely used for treating several illnesses in folk medicines. With increase in dose and time motility and sperm count decreased and increase in percentages of sperm debris, primordial cells and abnormal cells was observed (Obianime *et al*., 2010).

Piper longum *(*Long pepper*)* belongs to family Piperaceae. Piperine treatment at dose level of 10 mg causes decrease in weight of sex organs and testis. Study also revels damage to semiferous tubule, decrease in ledying cells, spermatids and spermatocytes desquamation (Malini *et al.*, 1999).

Piper nigrum (Black pepper) belongs to family Piperaceae. Mice treated for 90 days with 100 mg dose study shows negative effect on sperm parameters like fructose, sialic acid and litter size. Main effect was on seminiferous tubules i.e. loosing of germinal epithelium, giant cell occurrence, mixing o different stages of spermatids in spermatogenesis and intra epithelial vaccolation (Mishra and Singh, 2009).

Azadirachta indica (Neem) belongs to family Meliaceae. Neem flower and bark contraceptive efficacy in male albino rats was reported by Purohit *et al.*, 1990. Purohit., 1990 reported that neem leaf extract also has contraceptive efficacy in male rats.

Aphrodiasic and Infertility Treatment Potential

Leptadenia reticulata (Jivanti) belongs to family Apocynaceae. The chloroform extract of whole plant, revealed that plant has more potential of aphrodiasic activity compared to normal as it has got increased number of mounts, intromission interval, decreased latency of first mount as well as increase in post ejaculations time and number of ejaculations in both normal and stressed rats. Weight gain in testis (PL 0.05) prostate gland, seminal vesicle epididymis, vasdeferens, with decline in weight of adrenal gland indicates HPA/ HPG excess regulatory mechanism (Santosh *et al.*, 2011).

Fertility Regulation in Females

Estrous Cycle Disruptors

Rivea hypocrateriformis (Night Glory/ Midnapore creeper) belongs to family Convolvulaceae. Irregular estrous cycle with shortened metestrus and estrus and lengthened proestrus in reversible manner was induced in adult albino rats with the ethanolic extract of *Rivea hypocrateriformis* orally at dose of 200 and 400 mg/kg body weight. Extract showed antiovulatory effect as it declined graffian follicles number and increased atretic follicles (Shivalingappa *et al.*, 2002).

Momordica charantia (Bitter gaur/Karela) belongs to family Cucurbitaceae. Increased length of estrous cycle and irregular pattern of estrous cycle was observed in rats when administrated with oral dose of 25mg/100g body weight of methanolic extract of *Momordica charantia* seeds (Ifeanyi *et al.*, 2011).

Azadirachta indica (Neem) belongs to family Meliaceae. Irregular pattern of oestrous cycle with prolonged diestrus phase was observed in rats when administered with *a*lcoholic extract *Azadirachta indica* of flower at dose level of 1g/kg body weight produced (Gbotolorun *et al.*, 2008).

Azadirachta indica L. (Neem) in medicine ayurvedic system leaf of neem is used for fertility regulation. Reactive oxygen species (ROS) are induced by aqueous neem leaf extract (NLE) these reactive oxygen species does apoptosis of granulosa cells. Prior to ovulation for the achievement of meiotic competency of follicular oocytes require cellcycle proteins, survival factors and nutrients which are not available due to the apoptosis of the granulose cells. Both in follicular oocyte and granulosa cells apoptosis is mediated by the

NLE. Apoptosis by NLE is mediated by Mitrocondria and ROS generation. Detoration in quality of oocyte occurs due to apoptosis and due to this on mammals outcome is limited (Chaube *et al.*, 2014).

Jatropha gossypifolia (Bellyache Bush/Ratan jyoti) belongs to family Euphorbiaceae. The estrous cycle was prolonged with increase in duration of diestrus stage in female albino mice when administered with Ethanolic extract of *Jatropha gossypifolia* leaves at a dose level of 250 and 450 mg/kg body weight (Jain *et al.*, 2012).

Crocus sativus (Saffron, Kesar) belongs to family Iridaceae. It is a spice derived from flower is used for fertility control. Crude methanolic extract was tested for antifertility activity in female albino rat. Crude extract was administered at a dose of 1g/kg body weight/day for 12 days. Extract treated rat's estrous cycle become irregular which resulted in gestation failure. Restoration of estrous cycle was observed after recovery period but number of newborn pups was lesser than the control. The female reproduction in rats was effected by estrogenic/steroidal compounds present in *Crocus sativus* (Bashir *et al.*, 2020).

Anti-estrogenic Agents

Butea monosperma (Flame of the Forest/Dhak) belongs to family Fabaceae. Albino mice was treated with chloroform and Petroleum ether extract at dose level of 200 mg/kg body weight of *Butea monosperma* root resulted in estrogenic activity when given alone. When given along with ethinyl estradiol exhibited slight anti-estrogenic activity. There was increase in cholesterol level and weight of ovaries decreased after treatment with chloroform and Petroleum ether. These extracts also inhibit activity of G-6-PDH which indicates their anti-steroidogenic activity (Sharma *et a*l., 2012)

Cassia fistula (Golden Shower/Amaltas) belongs to family Caesalpiniaceae/ Fabaceae. Estrogenic activity was shown by female rats when treated with aqueous extract of *Cassia fistula* seeds but when given with estradiol valerate it prevented the estrogen-induced uterotrophic effect, this indicates anti-estrogenic nature with strong estrogen. It also prevented pregnancy when administered to mated females rats (Yadav *et al.*, 1999)

Bougainvillea spectabilis (Great Bougainvillea) belongs to family Nyctaginaceae. A significant decrease in estrogen level was observed when aqueous extract of *Bougainvillea spectabilis* leaves at a dose level of 300 mg/kg body weight was administered to Swiss albino mice (Mishra *et al.*, 2009).

Ocimum gratissimum (African Basil/Tulsi) belongs to family Lamiaceae. Female albino rats were administered with Acetone extract of *Ocimum gratissimum* stem alone at a dose level of 100 mg/kg body weight resulted in weak estrogenic activity but when provided with estradiol valerate exhibited antiestrogenic activity. Extract administration to mated female rats showed decline in live fetuses, fertility index and uterine implant number (Sripriya *et al.*, 2011).

Piper betle (Betel, Vine Betel, Betel Pepper) belongs to family piperaceae. It is widely used alternative for the synthetic drugs containing chemical for fertility regulation having side effects. Anti estrogenic activity was observed at the dose level of 500mg/kg body weight which was reflected in the form of significant increase in height of endometrium epithelium, uterus diameter and enodometrium thickness. No estrogenic activity was observed by the extract. APB exhibited significant antifertility activity 62.2% (Shah *et al.*, 2018).

Trigonella foenum graecum (Fenugreek) belongs to family Fabaceae. Extract of seeds of *Trigonella foenum graecum* were tested on female rats has antifertility and antiendrogenic activity. Biological components such as sialic acid, protein, glycogen and ascorbic acid decreased in uterus and ovary but cholesterol increased in uterus and ovary (Sharma and Bhinda, 2005).

Curcuma longa (Turmeric) and ***Carum carvi*** *belongs to family* Apiaceae. Extract of *Curcuma longa* rhizome *and Carum carvi* seeds were tested on female rats for 30 days. Oral administration of different doses to female albino rats showed antifertility activity. Hormonal assay of their study suggested increased estrogen levels in case of both drugs treated animals in ethanolic extract and also decreased LH (Luteinizing Hormone) and FSH (Follicle Stimulating Hormone) levels (Thakur *et al*., 2009).

Anti-Implantation Agents

Hibiscus rosa-sinensis (China Rose/Guhmohr) belongs to family Malvaceae. Anti-implantation activity was shown by rats when administered orally with Ethanolic extract of *Hibiscus rosa-sinensis* at a dose of 400mg/kg. endometrium thickening, increase in uterus diameter and weight of uterus was observed when treated with the extract (Vasudeva *et al*., 2008)

Calotropis procera (Sodom apple/Aak) belongs to family Asclepiadaceae. Strong anti-implantation activity was observed in female albino rats were administered with Ethanolic extract of *Calotropis procera* roots at the dose level of 250 mg/kg. Antiestrogenic activity was not detected in the extract (Ranab *et al*., 2002)

Terminalia bellirica (Baheda) is deciduous tree belonging to family Combertaceae. Ethanol and benzene extract of ***Terminalia bellirica*** bark were used for the antiimplantation property study on adult albino mice at dose of 25mg/10g body weight strong antiimplantation (inhibition 74.20%) activity was observed (Vishwanatha *et al*., 2009).

Allium cepa (**Onion**) belongs to family Alliaceae/Amaryllidaceae. The ethanolic extract of Allium cepa was found involved in inhibition of implant sites number at the dose level of 300 mg/kg though there was no change in ovulation. This study attributes towards antiimplantation propertiesof onion (Vishnu *et al*., 2003).

Moringa oleifera (Lam) belongs to family Moringaceae is used in Nigerian Traditional Medicine. Cold and hot aqueous extracts of *M. oleifera* leaves (MOL) was used to observe effect on the *in vitro* and *in vivo* uteroactive and contraceptive potentials. Uterine contractility with varying intensities was produced when administered *in vitro* with and hot aqueous extracts of MOL significantly ($p < .05$). 100% and 80% inhibition of conception was observed when administered with MOL cold aqueous extracts *in vivo* before and after mating, respectively. Hot aqueous extracts of MOL produced 96.6% and 58% effect *in vivo* (Attah *et al*., 2020).

Leptadenia reticulata (Jivanti) belongs to family Apocynaceae. Whole plant extract of *Leptadenia reticulata* showed ant implantation activity. Whole plant extract was prepared using ethanol and was given to albino rats for evaluation of anti-implantation and hormonal activities. Dose of 300mg/kg resulted in a strong anti-implantation (inhibition 100%) and uterotropic activity with no antiestrogenic activity (Rani *et al*., 2009).

Abortifacient Agents

Ailanthus excels (Indian Tree of Heaven/Ardu) belongs to family Simaroubaceae. Abortifacient activity was observed in Wistar albino rats were administered with hydroalcoholic extract of *Ailanthus excelsa* stem bark at a dose level of 400 mg/kg body weight showed abortifacient activity. Uterine weight was also increased by the extract. Antiestrogenic activity of extract was observed administered with ethinyl estradiol (Ravichandran *et al*., 2007).

Cannabis sativa (Hemp/Bhang) belongs to family Cannabaceae. Strong abortifacient activity was observed in female albino rats when administered with Alcoholic extract of *Cannabis sativa* leaves at dose level of 400 mg/kg body weight. Decrease in the uterine weight and ovarian was observed but body weight increased after treatment with extract. Serum progesterone level increased and serum estrogen level decreased. LH and FSH also found reduced (Zade *et al*., 2013).

Table 1: *Fertility regulation activity of different medicinal plants.*

S. No.	Name of the plant	Family	Common name	Plant parts used	Bioactive constituents	Fertility Regulation Activity
1.	*Martynia annua*	Martyniaccae	Devil's Claw	Root	Carbohydrates, alkaloids, flavonoid, tannins, terpenoids, steroids, phenol, cardiac and saponin glycosides.	Male Contraceptive
2.	*Maytenus emarginata*	Celastraceae	Kankhera	Leaf	Amino acids, fatty acids, tocopherols, sterols, glucosinolate and phenolic compounds.	Male Contraceptive
3.	*Opuntia elatior*	Cactaceae	Prickly Pear	Fruit	Alkaloids, polyphenols, phenolic and saponins.	Male Contraceptive
4.	*Nigella sativa*	Ranunculaceae	Blackseed or black cumin	Seed	Carbohydrates, fats, Vitamins, mineral elements, proteins, polyphenols flavonoids, alkaloids, steroids, terpenes coumarines, tannins and saponins.	Male Contraceptive
5.	*Terminalia chebula*	Combretaceae	Myrobalan, Hardad	Fruit	Lignin, phenolic acids, phytosterols, steroids, terpenoids, flavonoids, alkaloids, tannins, glycosides and carotenoids.	Male Contraceptive

6.	*Albizia lebbeck*	Fabaceae	Indian Siris	Bark	Alkaloids, saponins, tannins, flavonoids and coumarins.	Male Contraceptive
7.	*Aegle marmelos*	Rutaceae	Golden Apple, Bael	Leaf	Phenols, flavonoids, alkaloids, cardiac glycosides, saponins, terpenoids, steroids and tannins.	Male Contraceptive
8.	*Dactyloctenium aegyptium*	Poaceae	Crowfoot Grass, Beach Wiregrass	Whole Plant	Carbohydrates, proteins, amino acids, saponins, flavonoids, tannins, terpenoids and alkaloids.	Male Contraceptive
9.	*Butea monosperma*	Fabaceae	Flame of forest or palash	Leaf and Root	Alkaloids, butrin, butein, butin, isobutin, coreoopsin, monospermoside glucose, glycine, and an aromatic hydroxy compound.	Anti-estrogenic Agents and Male Contraceptive
10.	*Coccinia indica*	Cucurbitaceae	Ivy Gourd	Leaf	Linalool, beta caryoohyllene, alkaloids, flavonoids, saponin and phenol.	Male Contraceptive
11.	*Withania somnifera*	Solanaceae	Ashwagandha	Fruit	Phenols, flavonoids, tannins, saponins, alkaloids, steroids, terpenoids, glycosides, reducing sugars, volatile oils, zingerone and gingarols.	Male Contraceptive and infertility
12.	*Cuminum cyminum*	Apiaceae	Cumin	Seeds	Alkaloids, coumarin, anthraquinone, flavonoid, glycoside, protein, resin, saponin, tannin and steroid.	Male Contraceptive
13.	*Mentha arvensis*	Lamiaceae	Pudina	Leaf	Alkaloids, Steroids and Glycosides.	Male Contraceptive
14.	*Ocimum gratissimum*	Lamiaceae	Tulasi	Stem and Leaves	Alkaloids, phenolics, glycosides, resins, steroids, and tannins.	Anti-estrogenic Agents and Male Contraceptive

15.	*Piper longum*	Piperaceae	Long pepper	Seeds (Piperine)	Piperine, piperlongumine, sylvatin, sesamin, diaeudesmin piperlongumine, pipermonaline and piperurndecalidine.	Male Contraceptive
16.	*Piper nigrum*	Piperaceae	Black pepper	Fruit	Alkaloids, tannins, saponins, flavonoidscardiac glycosides and phenolic compounds.	Male Contraceptive
17.	*Azadirachta indica*	Meliaceae	Neem	Flower, Bark and Leaves	Steroids, triterpinoids, reducing sugars, alkaloids, phenolic compounds, flavonoids and tannins.	Estrous Cycle Disruptors and Male Contraceptive
18.	*Leptadenia reticulata*	Apocynaceae	Jivanti	Whole Plant	Stigma sterol, beta-sitosterol, flavonoids, pregnane glycosides, tocopherol and proteins.	Aphrodiasic and Anti-implantation
19.	*Rivea hypocrateri-formis*	Convolvu-laceae	Night Glory/ Midnapore creeper	Aerial Parts	Alkaloids, glycosides, saponins, tannins and phenolic compounds.	Estrous Cycle Disruptors
20.	*Momordica charantia*	Cucurbitaceae	Bitter gaur/ Karela	Seeds	Steroids, triterpinoids, reducing sugars, sugars, alkaloids, phenolic compounds, flavonoids and tannins.	Estrous Cycle Disruptors
21.	*Jatropha gossypifolia*	Euphorbiaceae	Bellyache Bush/Ratan jyoti	Leaves	Saponin, tannin, flavonoid, reducing sugars, cardiac glycosides, terpenoids, triterpenoids, steroids, xanthoprotein, and starch.	Estrous Cycle Disruptors
22.	*Crocus sativus*	Iridaceae	Saffron, Kesar	Flower	Anthocynins, flavonol glycosides, carotenoid and carboxaldehyde.	Estrous Cycle Disruptors

23.	*Cassia fistula*	Caesalpiniaceae/ Fabaceae	Golden Shower/ Amaltas	Seeds and Pods	Anthraquinone glycosides, flavonoids, phenolic compounds and carbohydrates.	Anti-estrogenic Agents
24.	*Bougainvillea spectabilis*	Nyctaginaceae	Great Bougainvillea	Leaves	Alkaloids, flavonoides, phlobatannins and terpenoids.	Anti-estrogenic Agents
25.	*Piper betle*	Piperaceae	Betel, Vine Betel, Betel Pepper	Petiole	Carbohydrates, alkaloids, gums, oils, steroids, glycosides, tannins, phenols, vitamins, organic acids and inorganic constituents.	Anti-estrogenic Agents
26.	*Trigonella foenum graecum*	Fabaceae	Dana Methi, Fenugreek	Seeds	Alkaloids, carbohydrates, glycosides, tannins, flavonoids, amino acids, proteins, mucilage and starch.	Anti-estrogenic Agents
27.	*Curcuma longa*	Zingiberaceae	Turmeric	Rhizome	Flavonoids, aminoacids, alkaloid, alkaloids, carbohydrates, phytosterols, fixed oils, fats, proteins, glycosides, saponins and tannins.	Anti-estrogenic Agents
28.	*Hibiscus rosa-sinensis*	Malvaceae	China Rose/ Guhmohr		Steroid, tannins, saponins and flavonoids.	Anti-Implantation Agents
29.	*Calotropis procera*	Asclepiadaceae	Sodom apple/ Aak	Root	Alkaloids, flavonoids, tannins, saponins and cardiac glycosides.	Anti-Implantation Agents
30.	*Terminalia bellirica*	Combertaceae	Baheda	Bark	Phytosterols, carbohydrates, flavanoids, phenolic compounds and tannins.	Anti-Implantation Agents
31.	*Allium cepa*	Alliaceae/ Amaryllidaceae	Onion	Bulb	Alkaloids, flavonoides, cardiac glycosides, terpenes and resins, Kampferol, β-sitosterol, ferulic acid and myritic acid.	Anti-Implantation Agents

32.	*Moringa oleifera*	Moringaceae	Lam	Leaves	Vanillin, omega fatty acids, carotenoids, ascorbates, tocopherols, beta-sitosterol, moringine, kaempferol and quercetin.	Anti-Implantation Agents
33.	*Ailanthus excels*	Simaroubaceae	Indian Tree of Heaven/ardu	Leaves, Stem and Bark	Alkaloids, glycosides, saponins, phenol, lignin, terpenoids, tannins and steroids.	Abortifacient agents
34.	*Cannabis sativa*	Cannabaceae	Hemp/Bhang	Leaves	Flavonoids, simple phenolics, alkaloids, steroids, saponins, terpenoids, tannins and reducing sugars.	Abortifacient agents

Martynia annua

Maytenus emarginata

Opuntia elatior

Nigella sativa

Terminalia chebula

Albizia lebbeck

Aegle marmelos

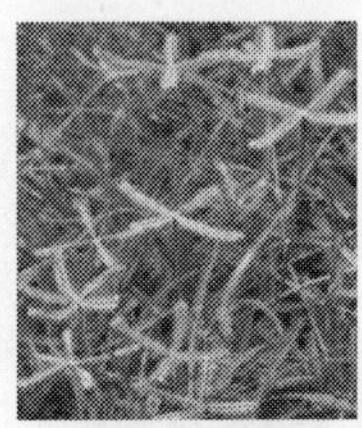
Dactyloctenium aegyptium

Butea monosperma

Coccinia indica

Withania somnifera

Cuminum cyminum

Mentha arvensis

Ocimum gratissimum

Piper longum

Piper nigrum

Fig. 1. *Pictures of the mechinical plants reviewed in this article*

CONCLUSION

In this review many research studies were included which reflects the use of different medicinal plant, tribal and traditional plants which are being used for regulating fertility around the world but still there is a lacuna between the studies conducted and their usable products available for human use. There are very few plant based medicines which are available in market that can be used for regulating fertility. There is need for evaluating such plants and their properties for different medicinal uses. Current need is to have a plant based drug product which is easily available with lesser side effects, low cost and have fertility regulating properties. Such product will be useful for population control and also in solving the in fertility related dysfunctions.

Most of the plant drugs are used traditionally without proper scientific knowledge as the same plant may have both the properties i.e. it may be used as contraceptive and it may also be used as a treatment for infertility. It has become important now to document medicinal plants which can be used by both male and female for treating reproductive health problems from our traditional knowledge present and also to analyse their pharmacological components and to explore secondary metabolites present in them. So it is need of hour that proper compound study with detailed analysis of plant's medicinal properties should be done and a proper formulated herbal product should be released for human use in market. There is need to study such properties of plant based compounds thoroughly on animals and then on human trait also. The researches present in above review also lack in the toxicity of the extract which are tested and also safety for human purpose.

It is evident from the review that further extensive studies and researches are required to evaluate the potential and exact effect of medicinal plants. Significant research is required for the evaluation of chemical and biological properties of less explored plants.

ACKNOWLEDGMENT

We are thankful to the Head and Department of zoology, Jai Narain Vyas University, Jodhpur, Rajasthan and UGC. Special thanks to Professor A.K. Purohit and Dr. R.P. Saran for their guidance. Heartfelt thank's to Dr. Kalpana Purohit (HOD, Department of English, Jai Narain Vyas University, Jodhpur, Rajasthan) for inspiring us to write this paper.

REFERENCES

- Adhikari, P.P., Talukar, S., and Borah, A. (2018). Ethenomedicobotanical study of indigenous knowledge on medicinal plants used for the treatment of reproductive problems in Nalbari district, Assam, India. *Journal of ethanopharmacology*. 210, 386-407.
- Allag, I.S., and Rangari, K. (2002). Extragenomic action of steroids on spermatozoa. Prospects for regulation of fertility. *Health Popul.* **25**(1):38-44.
- Attah, A.F.; Moody, J.O.; Sonibare, M.A.; Salahdeen, H.H.; Akindele, O.O.; Nnamani, P.O.; Diyaolu, O.A.; and Raji, Y. (2020). Aqueous extract of *Moringa oleifera* leaf used in Nigerian ethnomedicine alters conception and some pregnancy outcomes in Wistar rat. *South African Journal of Botany*. **129**, 255-262.
- Bashir, H.; Upadhyay, A,; and Mittal, D.K. (2020). Analysis of *crocus sativus* (saffron) on estrous cycle and induced reversible sterility in rat. *World Journal of Pharmacy And Pharmaceutical Sciences*. **9**(5), 1760-1765.
- Brinker, F. (1997). Inhibition of endocrine function by botanical agents, antigonadotropic activity. *Br J Phytother*. **4**:123-145.
- Chan, K. (2003). Some effects of toxic contaminations in herbal medicines. *Chemosphere*, **52**, 1361- 1371.

- Chaube, S.K., Shrivastav, T.G., Tiwari, M., Prasad, S., Tripathi, A., and Pandey, A.K. (2014). Neem (Azadirachta indica L.) leaf extract deteriorates oocyte quality by inducing ROS-mediated apoptosis in mammals. *SpringerPlus*. **3**, 464.
- Chauhan, A., Agarwal, M., Kushwaha, S., and Mutreja, A. (2007). Suppression of fertility in male albino rats following the administration of 50% ethanolic extract of *Aegle marmelos*. *Contraception*.**76**(6), 474-481.
- Chellaiah, M., Muniappen, A., Nagappen, R., and Savarimuthu, I. (2006). Medicinal plants used by traditional healers in kancheepuram districts of Tamil Nadu, India. *Ethanobiol Ethnomed.* **2**; 43-50.
- Dandapat, M.C., Mohapatro S.K., and Patro S.K. (1985). Management of subfertile males. *Indian Medical Gazette.* **1**:14.
- Dohle, G.R., Colpib, G.M., Hargreavec, T.B., Pappd, G.K., Jungwirthe, A., and Weidnerf, W. (2005). EAU Guidelines on Male Infertility. *European Urology*, **48**(5): 703-711.
- Gark, S.K., Mathur, V.S., Chaudhury, R.R. (1978). Screening of Indian plants for anti-fertility activity. *Indian J Exp Biol.* **16**(10):1077-1079.
- Gbotolorun, S.C., Osinubi, A.A., Noronha, C.C., and Okanlawon, A.O. (2008). Antifertility potential of Neem flower extract on adult female Sprague-Dawley rats. *Afr Health Sci.* **8**(3), 168–173.
- Gupta, R.S., Kachhawa, J.B.S., and Chaudhary, R. (2006). Antispermatogenic, antiandrogenic activities of *Albizia lebbeck* (L.) Benth bark extract in male albino rats. *Phytomedicine*. **13**(4), 277-283.
- Ifeanyi, A.C., Eboetse, Y.O., Ikechukwu, D.F., Adewale, O.A., Carmel, N.C., Olugbenga, O.A. (2011). Effect of *Momordica charántia* on estrous cycle of sprague-dawley rats. *Pacific Journal of Medical Sciences*. **8**(1).
- Jain, S., Choudhary, G.P., and Jain, D.K. (2012). Pharmacological evaluation of anti-fertility activity of ethanolic extract of *Jatropha gossypifolia* leaf in female albino mice. *Asian Pacific Journal of Tropical Biomedicine*. S1671-S1674.
- John, J.S. (1981). The continuing need for contraceptive research, *Fertil Steril.* **36**(6): 697-698.
- Kunwar, R.M., Nepal, B.K., Kshhetri, H.B., Rai, S.K., and Bussmann, R.W. (2006). Ethanomedicine in Himalaya: a case study from Dolpa, Humla, Jumla and Mustang districts of Neoal. *Journal of Ethanobiology and Ethnomedicine*, **2**, 27.
- Mali, P.C. (2013). Control of fertility in maleWistar rats treated with hydro alcoholic extract of *Withania somnifera*fruits. *IJPBS*. **7**, 13-21.
- Mali, P.C., A.S. Ansari, A.S., and Chaturvedi, M. (2002). Antifertility effect of chronically administered Martynia annua root extract on male rats. *Journal of Ethnopharmacology*. **82**, 61- 67.
- Malini, T., Manimaran, R.R., Arunkumar, J., Aruldhas M.M., and Govindarajulu, P. (1999). Effects of Piperine on testis of albino rats. *Journal of Ethnopharmacology*. **64**(3), 219-225.
- Malla, S.B., and Shakya, P.R. (1984). Medicinal plants of Nepal. In Nepal-Natures, Paradise. *White Lotus Ltd, Bangkok*, 261-297.

- Marles, R.J., and Farnsworth N.R. (1994). Plants as sources of antidiabetic agents. *Economic and medicinal plant research,* **6**, 149-187.
- Mishra, N., Joshi, S., Tendon,V.L., and Munjal, A. (2009). Evaluation of Anti-Fertility Potential of Aqueous Extract of *Bougainvillea spectabilis* leaves in Swiss albino mice. *International journal of Pharmaceutical sciences and drug research.* **1**(1), 19-23.
- Mishra, R.K., and Singh, S.K. (2009). Antispermatogenic and antifertility effects of fruits of *Piper nigrum* L. in mice. *Indian journal of Experimental Biology.* **47**, 706-714.
- Mohammad, M.A., Mohamad, M.M.J., and Dradk, H. (2009). Effects of Black Seeds (*Nigella Sativa)* on Spermatogenesis and Fertility of Male Albino Rats. *Research Journal of Medicine and Medical Sciences.* **4**(2), 386-390.
- Montaserti, A., Pourheydar, M., Khazaei, M., and Ghorbani, R. (2007). Anti-fertility effects of physalis alkekengi alcoholic extract in female rat. *Iran J Reprod Med.* **5(1):**13-16.
- Naik, B.S., Nim Bahadur Dangi, N.B., Sapkota, H.P., Wagle, N., Nagarjuna, S., Sankaranand, R., and Kumari, A. (2016). Phytochemical screening and evaluation of anti-fertility activity of Dactyloctenium aegyptium in male albino rats. *Asian Pacific Journal of Reproduction.* **5**(1), 51–57.
- Obianime, A.W., Aprioku, J.S., Chinagoro, T., and Esomonu, O. (2010). Antiferility effects of aqueous crude extract of *Ocimum gratissimum* L. leaves in male mice. *Journal of Medicinal Plants Research.* **4**(9), 809-816.
- Parchure, S.N. (1983). Charak Samhita. Sagar Publications. Pune. India.
- Prakash, A.O., Saxena, V., Shukla, S., and Mathur, R.(1985) Contraceptive potency of *Pueraria tuberosa* D.C. and its hormonal status. *Acta Eur Fertil.* **16**(1):59-65.
- Purohit, A. (1999). Contraceptive efficacy of neem leaves (*Azadirachta indica* A. Juss) in male rats. *Aryavaidan.* 12, 231-233.
- Purohit, A., Joshi, V.B., and Dixit, V.P. (1990). Contraceptive efficacy of *Azadirachta indica* (flower and bark) in male rats: Abiochemical and sperm dynamics analysis. *J. Curr. Biosci.*, **7**(4), 129-133.
- Ramya, M.C., Shivabasavaiah, and Shivanandappa, T. (2015). Reversible antifertility effect of *Opuntia elatior* Mill. fruit extract. *International Journal of Reproduction, Contraception, Obstetrics and Gynecology.* **4**(2), 392-397.
- Ranab, A.C., Kamatha., and Jagadish,V. (2002). Preliminary study on antifertility activity of *Calotropis procera* roots in female rats. *Fitoterapia.* **73**(1), 111-115.
- Rani, S., Manavalan, R., Kilimozhi, D., and Balamurugan, K. (2009). Preliminary study on anti–implantation activity of *Leptadenia reticulata* in female rats. *Int. J. PharmTech Res.***1**, 1403–1405.
- Ravichandran, V., Suresh, B., Sathishkumar, M.N., Elango, K., and Srinivasan, R. (2007) Antifertility activity of hydroalcoholic extract of *Ailanthus excelsa* (Roxb): An ethnomedicines used by tribals of Nilgiris region in Tamilnadu. *Journal of Ethnopharmacology.***112**, 189–191.
- Reddy, A.S., and Rajendran, S. (2018). Effect of methanolic leaf extract of *Butea monosperma* on fertility of male albino rats. *World Journal of Pharmacy and Pharmaceutical Sciences.* **7**(5), 1588-1597.

- Santosh B.T., Chitme, H.R., Rabbani, G., and Jafar M. (2011). Effect of *Leptadenia reticulate* linn on stress modulated sexual behavior of male rats, *Int. Res. J.pharm.***2**(10), 27-36.
- Shah, S.K., and Jhade, D.N. (2018). Evaluation of antifertility potential of *Piper betle* (Petiole) on female wistar rats "rising approaches of herbal contraception" *Biochemistry and Biophysics Reports.***15**, 97-102.
- Shaik, A., Kanhere, R.S., Cuddapah, R., Nelson, K.S., Vara, P.R., and Sibyala, S. (2014). Antifertility activity of *Artemisia vulgaris* leaves on female wistar rats. *Chin J Nat Med.* **12**(3):180-185.
- Shaik, A., Yalavarthi, P.R., and Bannoth, C.K. (2017). Role of anti fertility medicinal plants on male and female reproduction. *Journal of Complementary and Alternative Medical Research.* **3**(2), 1-22.
- Sharm, S.K., Rai, G., Vasudeva, N. (2012). Anti-fertility Investigation of *Butea monosperma* (Lam.) Kuntze Root in Female Albino Mice. *Research Journal of Medicinal Plant.* **6**(3), 260-266.
- Sharma, D.K., and Mali, P.C. (2016). Inhibition of spermatogenesis with the treatment of 50% methanolic extract of *maytenus emarginata* leaves in albino rats. *International Journal of Pharmaceutical Sciences and research.* **7**(8), 3379-3387.
- Sharma, J.D., and Bhinda, A. (2005). Antifertility activity of Steroidal extract of *Trigonella foenumgraceum* (seeds) in female rats. *Asian J.Exp.Sci.* **19**(1), 115-120.
- Sharma, N., and Jacob, D. (2001). Antifertility investigation and toxicological screening of the petroleum ether extract of the leaves of *Mentha arevensis* L. in male albino mice. *Journal of Ethnopharmacology.* **75**(1), 5-12.
- Sharma, P.V., (1993). Dravya Guna Vigyan, Chaukhamba Bharati Academy.n Varansi. India.
- Shivalingappa, H., Satyanarayan, N.D., Purohit, M.G., Sharanabasappa, A., Patil, S.B. (2002) Effect of ethanol extract of *Rivea hypocrateriformis* on the estrous cycle of the rat. *Journal of Ethnopharmacology*. **82**(1), 11–17.
- Sripriya, S., Yuvaraj, G., Nema, R.K., Kumar,V.M.; and Deecaraman, M. (2011). Evaluation of Antifertility activity from Stem Part of *Ocimum gratissimum* in Acetone extracts. *International Journal of Pharmaceutical and Clinical Research.* **3**(2), 41-44.
- Srivastav, A., Chandra, A., Singh, M., Jamal, F., Rastogi, P., Rajendran, S.M., Bansode, F.W., and VijaiLakshmi. (2010). Inhibition of hyaluronidase activity of human and rat spermatozoa *in vitro* and antispermatogenic activity in rats *in vivo* **by** *Terminalia chebula*, a flavonoid rich plant. *Reproductive Toxicology.* **29**(2), 214-224.
- Thakur, S., Bawara, B., Dubey, A., Nandhini, D., Chauhan, N.S., and Saraf, D.K. (2009). Effect of *Carum carvi* and *Curcuma longa* on hormonal and reproductive parameter of female rats. *Int.J.Phytomedicine.***1**, 31-38.
- United nations World population prospects. The revision. Executive summary, department of economic and social affairs population division. New York. 2007; 1-21.
- Vaidya, P., Padmashali, S., Vagdevi, H.M., and Sathyanarayana, N.D. (2006). Antifertility effect of the plant *Balanites roxburghii* (Balanitaceae) in female rats. *Indian J Pharm Sci.* **68**(3):347-351.

- Vasudeva, N., and Sharma, S.K. (2008). Post-Coital Antifertility Activity of *Hibiscus rosa-sinensis* Linn. roots. *Evid Based Complement Alternat Med.* **5**(1), 91–94.
- Venkatesh, V., Sharma, J.D., and Raka Kamal. (2002). A comparative study of effect of Alcoholic Extract of *Sapindus emarginatus, Terminalia belerica, Cuminum cyminum* and *Allium cepa* on Reproductive organs of male Albino rats. *Asian.J. Exp.Sci.* **16**(1&2), 51-63.
- Verma, H.P., and Singh, S.K. (2017). Antifertility efficacy of *Coccinia indica* in male mice and its possible mechanisms of spermatogenesis. *General and Comparative Endocrynology*. **241**, 89-99.
- Vishnu, N., Thakare., Pankaj, S., Dhote, V., and Avinash, D. (2003). Antifertility activity of Ethanolic extract of *Allium cepa* Linn in rats. *Int.J.Pharm Tech Res*. **1**(1), 73-78.
- Vishwanatha, T., Satishagouda, S., Patil, J.S., and Patil, B.S. (2009). Anti-implantation activity of *Terminalia bellirica* bark extracts in female albino rats. *BioTechnology.* **3**(4), 257260.
- Williamson, E.M., Okpako, D.T., and Evans, F.J. (1996). Pharmacological methods in phytotherapy research: Selection preparation and pharmacological evaluation of plant material. John Wiley and Sons Ltd., London. **1**:191-212.
- Yadav, R., and Jain, G.C. (1999). Antifertility effect of aqueous extract of seeds of *Cassia fistula* in female rats. *AdvContracept*. **15**(4), 293-301.
- Yakubu, M.T., Oladiji, A.T., and Akanji, M.A. (2007). Male sexual dysfunction and method used in assessing medicinal plants with aphrodisiac potentials. *Pharmacognosy Reviews* **1**: 49-56.
- Zade, V., Wikhe, M., Dabhadkar, D., Dawada, S., and Patil, U. (2013). Antifertility efficacy of *Cannabis sativa* leaves on female albino rats. *IJSIT* . **2**(2), 107-117.

CHAPTER

12

Combating Diabetes with Green Medicines—A Review

Manisha Solanki and Ashok Purohit
Department of zoology, J.N.V. University,
Jodhpur (Rajasthan), India. 342001
e-mail: manusolanki999@gmail.com, purohitashok14@gmail.com

ABSTRACT

Diabetes is one of the major chronic diseases. It is characterized by elevated blood glucose level with many health complications related to a group of metabolic disorders. There are a number of oral hypoglycemic agents and drugs are available in the market but prolonged intake of these drugs cause adverse effects on health like organ damage, low blood sugar level, nausea, stomach problems etc. so there is a strong need to find an alternative mode of medication approach such explain in Ayurveda, Sidha and traditional therapies for the treatment of diabetes. There are a number of plants are described in Ayurveda to treat several diseases and health problem including diabetes. The Great Indian Thar Desert has a number of plants with antihyperglycemic activities. Western Rajasthan has harsh climatic condition and scarcity of water. In spite of all adverse conditions, still this area has a number of medicinal plants. The present study focuses on various plants of Western Rajasthan which shows antidiabetic properties.

Key Words: Antidiabetic, Chronic disease, Traditional, Ayurveda, Herbal

INTRODUCTION

In this world every person wants a happy and healthy life. Every person knows that the death is unalterable and they have to face it one day but no one wants an ill life and a dolorous death. Chronic diseases do not cause a sudden death rather that they cause illness progressively and at the end disease become cureless. There are five types of chronic diseases around the world, which are heart diseases, stroke, cancer, chronic respiratory diseases and diabetes. Four out of five are cause death of people and numbers of people killed by these chronic diseases are much greater than any other infectious diseases and health problems. The impact of these chronic diseases is more in low and middle income countries (WHO, 2005).

Out of four diabetes mellitus is an atrocious chronic disease which worldwide known as diabetes, is a group of metabolic disorders which is characterized by enhanced level of blood glucose within a time period and associate with the insufficient production of insulin, produced by beta cells (β-cells) (Heydemann, 2016).The function of insulin is to get glucose into the cells of our body (Patel *et al.*, 2012).

Diabetes is affecting more than 400 million people all over the world. Its dispersal is estimated to grow to 642 million by 2040, and it is expected to become the 8th leading cause of death by 2030 (Ghoul *et al.*, 2017). Diabetes is affecting human being since ancient time and today it is recognizes as a major health problem. According to WHO first global report on diabetes today it is one of the major health issues which is affecting people all around the world. According to the International Diabetes Federation (IFD) and World Health Organization (WHO) today diabetes is a major cause of death. More than 425 million people of the world's population are suffering from this disease in 2017. These data are much higher than the previous estimated 382 million people in 2013 and now expected to rise up to 693 million in 2045 (Broschag *et al.*, 2016; International Diabetes Federation, 2017). In Asian population the incidence of diabetes has raised up rapidly in recent decades. In 2007, more than 110 million people were suffering from diabetes in Asia. That is why prevention and proper control should be on top public health priority in Asian population (Chan *et al.*, 2009).

India holds first rank in top five diabetic countries in the world and so referred as diabetic capital of the world according to Times of India which is especially due to less awareness, knowledge and consciousness among people (Singla *et al.*, 2019). There are 3 general types of diabetes mellitus- Type I Diabetes mellitus (IDDM), Type II (NIDDM) and Gestational diabetes mellitus (GDM). Out of these 2 are especially responsible for death risk.

Type I DM or Insulin Dependent DM- In this type of diabetes pancreatic β-cells lose their capacity to produce insulin (Ferizi *et al.,* 2018). In most of the cases this form is found in juvenile, so it is also known as juvenile diabetes. Its symptoms can appear at any stage of life (WHO, 1999). People with type 1 diabetes are totally dependent upon external source of insulin for their survival. The therapy for type 1 diabetes includes measurement of blood glucose level time to time (Diabetic Care, 2010).

Type II DM or Non Insulin Dependent DM- Diabetes mellitus occurs all over the world but the majority of occurrence of T2DM is most common in developed countries (Broschag *et al.*, 2016). In T2DM the problem is that there is insulin deficiency or insulin resistance. Along this T2DM is more common than type 1 and the exposure is more in those people who are middle aged or have obesity (Kalofutis *et al.*, 2007). T2DM was remaining undiagnosed for years, the reason is that hyperglycemia develops slowly and at earlier stage it does not shows the classic symptoms of diabetes. People with T2DM have more risk of macro-vascular and micro-vascular complications. The reason of increasing expansion of obesity and type-2 diabetes in children is the effect of globalization and industrialization in societies (Zimmet, 2000).

Fatty diet, migration of people to big cities (urbanization), unhealthy life style, lack of physical exercise, stress, obesity are some of the major factors behind increasing the risk of having diabetes type 2 among younger population as compare to western countries (Benerjee *et al.*, 2020). The development of diabetes is associated with several pathogenic processes. Like the autoimmune destruction of the β-cells of pancreas which leads to insulin deficiency to an abnormal range that result in resistance to insulin action (Samuel *et al.*, 2012).

Type 1diabetes also has genetic reasons and inherited with multiple genes having some HLA genotype which influence the risk of diabetes. In people who are high strung genetically, diabetes can be triggered by a number of environment factors for e.g., viral infection or diet. Type 2 diabetes happens primarily due to daily life style factors and genetics. There are so many factors which are involve in the development of type 2diabetes e.g. obesity (body mass index of greater than thirty), lack of physical activity, unhealthy diet, stress, tension, and urbanization. There is more risk of getting hyperglycemia in people who have increased body mass index (BMI) and over waist. Excepting these reasons there are some more other reasons which increase the risk of diabetes e.g. intake of sugar-sweetened beverages, saturated fatty acid, high fat food intake and enough consumption of dietary fibers. They associate with increasing the body weight and so they linked with having T2DM (Ley *et al.*, 2014; Imamura *et al.*, 2015).

Diagnosis methods and criteria have been updated over the time. According to World Health Organization (WHO), diabetes is condition which identified by observing increased blood glucose level.. According to the International Diabetes Federation the range of normal glucose level in blood is considered between 72mg/dl to 110mg/dl. Even after 2 hours of postprandial condition it will reach only 120mg/dl in normal condition.

DM type 2 has affected high population. People with diabetes have 25 times more risk of blindness, 17 times more risk of kidney disease, 30 to 40 times more risk of amputation, 2 to 4 times risk to myocardial infarction. T2DM a major health issue is associated with morbility, death risk and health care cost. The science of T2DM is complex and has bad effects on various ogans like liver, skeletal muscle, adipocytes, gut and beta cells (Bener *et al.*, 2018). The risk of cardiovascular mobility and mortality is increased by the abnormal blood glucose level (Tziakas *et al.*, 2005). People with high blood pressure have more chances of diabetes. Along this elevated blood pressure can lead to many diabetic complications (Kim *et al.*, 2006). One of the most common complications of diabetes is diabetic neuropathy which draws up a wide variety of clinical syndromes representing a complex interaction with pathogenic factors (Melendez-Ramierz *et al.*, 2010). The appearance of high proteinuria and the nephritic syndrome is common in this concern and this leads to End Stage Renal Disease (ESRD) (Tridgell *et al.*, 2010).

Diabetes type 1 is characterized by reduced level of insulin and can be cure with the insulin treatment in which patient require daily administration of insulin to lowers the blood glucose level (Okur *et al.*, 2017). There are a number of side effects that are associate with the use of the synthetic insulin for example swelling of arms and legs, weight gain, low blood glucose (hypoglycemia), chances of heart failure, blurred vision, increased heart rate, tingling in hands, feet, lips or tongue, headache and lack of sleep.

Diabetes type 2 is characterized by high blood glucose level, dyslipidemia and increased insulin resistance which leads to diabetes associated health complications (Benerjee *et al.*, 2020). There are many synthetic medicines are available for treatment of diabetes which controls the condition of hyperglycemia and they can be taken oral as well as intravenous for example Biguamides (Metformin), Sulphonyl ureas (SUs)., Meglitinides, Alpha-Glucosidase Inhibitors (AGIs), Thiazolidinediones (TZDs), Incretin Hormones, Dipeptidyl Peptidase-IV (DPP-IV), Glucagon-like Peptide-1 (GLP-1), Sodium -Glucose co- Transporter 2(SGLT2) Inhibitors. But all these drugs have very bad effect on patient's health. Sooner or letter they damage other organs and cause many health complications like term use of metformin can cause, Vitamin B12 defiency, lactic acidosis with renal impairment, nausea, anorexia and diarrhea (Jackson *et al.*, 1987). SUs cause weight gain, hypoglycemia, renal impairment and liver cirrhoses. The use of AGIs cause bloating, abdominal discomfort, diarrhea and flatulence (Chiasson *et al.*, 2002).

In the plant kingdom there are abundance and richness of many remedies which can prevent, control and even cure many human health problems and also for maintenance of various health and normal physiological activity. On the global level the role of traditional medicines for solution of health problems is invaluable. This is a shocking fact that about 80% of the people who are living in less developed county like India are depends on traditional medicine for their health care needs (World Health Organization, 2010).

In present scenario medical world is looking for an alternative treatment method to cure this evil without causing other harm to mankind not only from health point of view but also form financial backbone of a country and the search brings every one close to the nature.

Plants are integral part of ancient medicinal system "Ayurveda". The principles of ayurveda are compatible with the modern time medicinal system. According to some evidences ayurveda have some health care therapies which provide relief from many chronic diseases like musculo-skeleton disorder, osteoarthritis, diabetes mellitus and many more (Sharma and Agarwal, 2015). In the plant kingdom there are abundance and richness of many remedies which can prevent and cure many human health problems and also for maintenance of various health and normal physiological activity. On the global level the role of traditional medicines for solution of health problems is invaluable. This is a shocking fact that about 80% of the people who are living in less developed county like India are depends on traditional medicine for their health care needs (World Health Organization, 2010). Herbal plants have potential to cure and prevent many diabetic complications with less side effects and a lower economic cost (Parera *et al.*, 2017). In last few years the field of herbal medicines are becoming more popular in both developed and developing countries, the reason is that they have natural origin and have less side effects on human health (Grover *et al.*, 2002).

India has a dynamic variety of forest, mountains, plains, rivers, flora and fauna. Every kind of vegetation is found in different climatic and geographical regions of India. Rajasthan is the second largest state of India having harsh and diverse climatic condition. In west The Great Indian Thar Desert is situated where in east plains and in south east plateau. The great Arawali hills also contribute to divide Rajasthan into two different zones. Rajasthan has a variety of plants and almost all are medicinally useful and being used to treat many health problems and diseases (Shikha *et al.*, 2011).

There are some of the plants are described which are helpful for treatment of Diabetes mellitus:

1. ***Acacia arabica:*** Common name- Babooln, Family - Fabaceae

   The occurrence of this plant is in all the parts of India in wild habitat. The plant extracts use as an antidiabetic agent by the acting as secretagouge to release insulin. it controls the hyperglycemic condition in control rates but not in alloxanized animals. Powered seeds help to initiate release insulin from pancreatic beta cells when administrated in rabbits (Wadood *et al.*, 1989).

3. ***Allium cepa:*** Common name- Onion, Family- Amaryllidaceae

   Study says that onion has very effective potential of skin disease like dermal healing in diabetic patients, with experimental mechanism of action related to tensile strength improvement and anti-diabetic effect (David *et al.*, 2015).

4. ***Allium sativum:*** Common name- Garlic, Lasasun, Family- Amaryllidaceae

   There is decrease in high blood glucose level when diabetic rabbits are treated with garlic (Grover *et al.*, 2002).

5. ***Aegle marmelose:*** Common name- Bael, Sirphal, Holy fruit tree, Family- Rutaceae

   There is effective control on blood glucose, urea, body weight, liver glycogen and serum cholesterol in alloxanized rats (60 mg/kg IV) when treated withaqueous extract of *Aegle marmelose* (Grover *et al.*, 2002).

6. ***Aloe vera or Aloe barbadensis:*** Common name- Ghee Kunwar, Kumar panthu, Family- Asphodelaceae

   Effective results are gain by clinical experiments that the *Aloe vera* gel is reducing increased blood glucose and Hb1Ac (Pradeep *et al.*, 2016).

7. ***Artemisia pallens:*** Common name- Davana, Family- Asteraceae

   This plant is endemic in south India and traditionally used for treatment of diabetes mellitus. Anti-hyperglycemic effects are showed in glucose feed hyperglycemic rats and alloxanized rats when treated with methanol extract of aerial parts of this plant (Grover *et al.*, 2002)

8. ***Azadirachta indica:*** Common name- Nim or Neem, Family- Meliaceae

   The plant extract is shows pharmacological activity independent of its time that either it is given prior or after treating with alloxan (Khosla *et al.*, 2000).

9. ***Beta vulgaris:*** Common name- Chukkander, Garden beet, Family- Amaranthaceae

   This plant contains many glycosides (e.g. beta vulgarosides II, III, and IV). These glycosides increase glucose tolerance in OGTT conducted in rats. Along this extract of this plant inhibit non-enzymatic glycosylation of skin proteins in STZ diabetic rats (Grover *et al.*, 2002).

10. ***Bougainvillea spectabilis:*** Common name-great bougainvillea, paper flower plant, Family- Nyctaginaceae

    Roots of this plant shows antiviral and antibacterial properties and leaves have antihyperglycemic activity. Bark extract of this plant is effective on albino rats (Purohit and Sharma, 2006).

12. ***Cappparis deciduas:*** Common name- Kurel, Kair or Pinju and Caper plant, Family- Capparaceae

    It is very important plant of Thar Desert and has very important medicinal values. Significant hypoglycemic effects are examined in alloxanized rats. This plant also shows anti-oxidant and hypolipidemic activity (Yadav *et al.*, 1997; Grover *et al.*, 2002).

13. ***Citrullus colosynthis:*** Common name- Bitter apple, bitter cucumber, desert gourd or makkal

    Family- Cucurbitaceae

    Along with diabetes, dried pulp of Citrullus is also use to treat constipation, edema, bacterial infection and cancer. In experimental diabetic rats, aqueous extract shows anti-hyperglycemic activity (Gurudeeban and Ramanathan, 2010).

14. ***Coccinia indica:*** Common name- Scarlet gourd, tindora and kowai fruit, Family- Cucurbitaceae

    There is significant hypoglycemic effect showed in normal rats when treated with pectin isolated from *C. indica* and this action is due to stimulation of glycogen synthetase activity and decrease of phosphorylase activity (Kumar *et al.*, 1993; Grover *et al.*, 2002).

15. ***Cyamopsis tetragonoloba:*** Common name- Guar, cluster bean or Guvar bean, Family- Fabaceae

    Leafs and fruits are used in many human diseases. Methanolic extract of fruit significantly reduce blood glucose level of STZ-induced diabetic rats (Kumareswari 2018).

17. ***Ficus religosa:*** Common name- holy pipal, bodhi tree or pippala tree, Family- Moraceae

    Aqueous extract of plant shows significant reduction in STZ-induced diabetic rats (Pandit *et al.*, 2010).

18. ***Gymnema sylvestre:*** Common name- Gudmar or Merasingi and Periploca of the woods, Family- Apocynaceae

    Methanolic extract of*Gymnema sylvestre* shows significant blood glucose reduction in diabetic rats and also have antioxidant (Kumar *et al.*, 2017).

20. ***Lantana camara:*** Common name- Caturang or Ghaneri, Family- Verbenaceae

    A new stearoyl glucoside of ursolic acid significanlty lowers the blood glucose level in STZ-induced diabetic rats (Kazmi *et al.*, 2012).

21. ***Momordica charantia:*** Common name- Bitter gourd, Karela, Balsam, Family- Cucurbitaceae

    This plant is also called "vegetable insulin", because its componants have similar structure like animal insulin. This plant have very extensive anti-diabetic properties and helpful in preventing diabetic complications. The hypoglycemic activity is experimentally have been proved on various experimental animals (Basenet *et al.*, 2000). According to some experiments this plant reduces the blood glucose level by suppress the expression of pro-inflammatory cytokines like IL-6 and IFNγ (Agustin and Rifa, 2015).

24. ***Phyllanthus niruri:*** Common name- Jangali Amla, Family- Phyllanthaceae

    This plant has diuretic, hypotensive and hypoglycemic properties. Regular dose showed significant reduction in blood glucose and systolic blood pressure (Grover, *et al.*, 2002).

26. ***Salvadora oleoides:*** Common name- Bada peelu, Family- Salvadoracee

    This plant is rich in sulphur containing compound and these sulphure compounds carries hypoglycemin properties. Ethanolic extract of leaves, flower and seeds significantly reduced high blood glucose level and increase insulin secretion from pancreatic β-cells (Yadav *et al.*, 2008).

27. ***Trigonella foenum-graecum:*** Common name- Fenurgreek or Methi. Family- Fabaceae

This plant has significant antihyperglycemic activity. It works on k-ATP channals and stimulate Ca^{++} influx, so as result blood glucose level slow down (Geil and Shane-McWortes, 2008).Seeds of this plant reduce both fasting sugar and postprandial blood sugar level without causing any known harm (Bagchi, 2016; Goyal, 2015).

28. ***Tinospora cordifolia:*** Common name- Heart-leaved moonseed, Guduchi or Amarta, Family- Menispermaceae

Oral administration of aqueous extract of root in alloxan induced rats showed a effective decrease in blood glucose level, brain lipid level, hepatic glucose-6-phosphatase, serum acid phosphatase, alkaline and lactate dehydrogenase but there is increase in body weight, hepatic hexokinase also seen (Stanley *et al.*, 1999).

CONCLUSION

Chronic diseases put a very bad impact on people's health but also a country's economy and their man power. Many scientists, researchers and institutes are searching for an effective herbal formulation which can cure these diseases. Herbal plants are a huge source of modern medicine because they possess a number of bioactive compounds. In this series the laboratory of Prof. Ashok Purohit at Zoology department, Jai Narain Vyas University, Jodhpur plays and significant role in searching importance of many medicinal plant in a way to cure diabetes, atherosclerosis and anti-fertility. Under the supervision of Dr. Ashok Purohit a significant work was also done on different ethological and environmental aspects of Bats and vultures. Many plants are investigated under the supervision of Dr. Ashok Purohit e.g. Neem (*Azadirachta indica*), Neem seed oil, kitchen spices (Coriender, Cumin, Turmeric), and Thar Desert Vegetation. Research work was also done on various kinds of diabetes mellitus.

Hyperglycemic condition was successfully controlled by Purohit *et al.,* (2008) by administration of *Embilica ribis* berries extract on alloxan induced diabetic rats. Also a significant work was done by Purohit and Daradka (2000) on anti-diabetic activity of *Syzygium cumini* seed extract in alloxan induced diabetic rats. Dixit *et al.,* (1992) has also use Neem (*Azadirachta indica*) to lowers blood glucose level successfully.

Ethanolic extracts administration of various parts of *Tecomella undulata* (leaf, bark, root and flower) shows a significant reduction in serum glucose level and shows regeneration in pancreas in STZ-indused diabetic albino rats (Lal *et al.*, 2017).

Various extracts administration of various extract of *Bougainvillea spectabilis* and *Capparis decidua* in streptozotocin-induced diabetic rats shows a significant reduction in increased blood glucose level (Purohit and Sharma, 2006)

Ethanolic extract administration of *Leptadenia pyrotechnica* stem shows and significant decrease in serum glucose level regeneration in pancreatic tissue in dexamethasone-induced diabetic rats (Solanki and Purohit, 2021).

ACKNOWLEDGEMENT

First of all I would like to thank my mentor and research supervisor Prof. Ashok K. Purohit, Former Head, Department of Zoology, for his continual inspiration, exemplary guidance and support to my work. I also convey my thanks to the Head of Department of Zoology for providing all kind of facilities I needed. I owe my special and sincere thanks to Prof.

Kalpana Purohit for her encouragement and support. I admire the support of Dr. R P Saran and Dr. H Ram Barad for helping me. I grateful to CSIR, New Delhi for providing me financial support.

REFERENCES

- Agustin F. and Rifa M. (2015): Efectivity combination of bitter melon (*Momordica charantia*) and *Andrographis paniculata* extract to suppress proinflam matory cytokinesis in diabetes mouse models. *J. Biotropika*. 3 (2): 81-85.
- Bagchi D., Swaroop A., Preuss H.G., Bagchi M. and Kumar P. (2016): Antidiabetic efficacy of a novel fenugreek seed (*Trigonella foenum-graecum, Fenfuro*) extract in patients with Type-2 diabetes. *Int J Med Scie*. 12 (10): 825-831.
- Bener A., Eliacik M., Cincik H., Ozturk M., DeFronzo R.A. and Ghani M.A. (2018): The input of vitamin D deficiency on retinopathy and hearing loss among type-2 diabetic patients. *Bio Med Res. Inter.* Article ID 2714590.
- Benerjee J., Dhas Y. and Mishra N. (2020): Middle-aged Indians with Type-2 diabetes are at higher risk of biological ageing with special reference to serum CDKN2A. *Journal of Diabetic research*. Vol. 2020. 10 pgs.
- Broschag M., He Y.L., Kjems L. and Foley J. (2016): Diabetes Ahead of Print, published online April 5, 2016: 1-30.
- Chaisson J.L., Josse R.G., Gomis R., Hanefeld M., Karasik A. and Laakso M. (2002): Acarbose for prevention of type-2 diabetes mellitus: the STOP-NIDDM randomized trial. *Lancet*. 359: 2072-7.
- Chan J.C., Malik V., Jia W., Kadowaki T., Yajnik C.S., Yoon K.H. and Hu F.B. (2009): Diabetes in Asia: epidemiology, risk factors and pathophysiology. *JAMA*. 301, 2129-2140.
- David E.T., Mendimi N.J., Tatsimo S., Nnanga N., Edmond J. and Miatto D. (2015): Effect of A methanol extract of *Allium cepa* Linn. On incisional wound healing in alloxan induced diabetic mice. *Applied Med Res*. 1(3): 90-93.
- Diabetic Care (2010): Diagnosis and classification of diabetes mellitus. American Diabetes Association Position Statement. *Diab Care*. 33(SI): S61-S69.
- Dixit V.P., Jain P. and Purohit A.K. (1992): Medicinal uses of Neem (*Azadirachta indica*) in fertility regulation, Diabetes and Atherosclerosis. *Recent Advanc in Medici Arom and Spice Crops*. 2: 463-471.
- Ferizi L., Dragidella F., Spahiu L., Begzati A. and Kotori V. (2018): The influence of type-1 diabetes mellitus on dental caries and salivary composition. *Int J Dentistry*.
- Ghoul El.B., Daaboul Y., Korjian S., Alam El. A., Manour A., Hariri E., Samad S., Salameh P., Dahdah G., Blacher J., Safar M.E. and Aoun Bahous S. (2017): Etiology of end-stage renal disease and arterial stiffness among hemodialysis patients. *Biomed Res Int*. 2017.
- Goyal M. (2015): Traditional plants used for the treatment of diabetes mellitus in Sursagar constituency, Jodhpur, Rajasthan- An ethno medicinal survey. *J Ethnoph*. 174: 364-368.
- Geil P. and Shane, M.C., Whortes. (2008). Dietary supplements in the management of diabetes potential risks and benefits. *J Am Diet Assoc*. 108(4):559-565

- Grover J.K., Yadav S. and Vats V. (2002): Medicinal plants of India with anti-diabetic potential. *J Ethnopharmacology*. 81: 81-100
- Gurudeeban S. and Ramanathan T. (2010): Anti-diabetic effect of *Citrullus colocynthis* in alloxan induced diabetic rats. *Inventi J.* 1(1): 1-4.
- Heydemann A. (2016): An overview of murin high fat diet as a model for type-2 diabetes mellitus. *J Diabetes Res.*
- Imamura F., O'Connor L., Ye Z., Mursu J., Hayashine Y., Bhupathiraju S.N. and Forouhi N.G. (2015): Consumption of sugar sweetened beverages, artificially sweetened beverages, and fruit juice and incidence of type-2 diabetes: systematic review, meta-analysis and estimation of population attributable fraction. *Brit Med* J. 351: h3576.
- International Diabetes Federation (2017): IDF DIABETES ATLAS- Eighth Edition.
- Jackson R.A., Hawa M.I., Jaspan I.B., Sim B.M., Disilvio L., Featherbe D. and Kurtz A.B. (1987): Mechanism of metformin action in non-insulin-dependent diabetes. *Diabetes. 36*: 632-40.
- Kalofutis C., Piperi C., Kalofoutis A., Harris F., Phoenix D. and Singh J. (2007): Type II diabetes mellitus and cardiovascular risk factors: current therapeutic approaches. *Exp Clin Cardiol Spring*; 12 (1): 17-28.
- Kazmi I., Rahman M., Afzal M., Gupta G., Saleem S., Afzal O., Shanaryar M., Nautiyal U., Ahmed S. and Anwar F. (2012): Antidiabetic potential of urosolic acid steroyl glucoside: A new triterpenic gycosidic ester from *Lantana camara. Filoterpia.* 83(1): 142-146.
- Khosla P., Bhanwara S., Singh J., Seth S. and Srivastava R.K. (2000): A study of hypoglycemic effects of *Azadirechta indica* (Neem) in normal and alloxan diabetic rabbits. *Indian J. of Physiology and Pharmacology* 44 (1): 69-74.
- Kim S.H., Lee S.J., Kang E.S., Kang S., Hur K.Y., Lee H.J., Ahn C.W., Cha B.S., Yoo J.S. and Lee H.C. (2006): Effects of lifestyle modification on metabolic parameters and carotid intima media thickness in patients with type-2 diabetes mellitus. *Metabolism.* 55 (8): 1046-1052.
- Kumar P., Rani S., Arunjyoti B., Chakrapani P. and Rojarani A. (2017): Evaluation of antidiabetic activity of *Gymnema sylvestre* and *Androgyrophis paniculata* in Streptozotocin-induced diabetic rats. *Int J of Pharmacognosy and Phytochemical Research.* 9(1): 22-25.
- Kumarswari T. (2018): Antidiabetic activity of foliar application of extract of *Ulva reticulata* grown fruit of *Cyamopsis tetragonoloba* (L.) taub in alloxan-induced diabetic rats. *J of AppliedBiotechnology & Bioengineering.* 5(3): 146-149.
- Kumar G.P., Sudheesh S. and Vijayalakshmi N.R. (1993): Hypoglycemic effect of *Coccinia Indica*: mechanism of action. *Planta Medica.* 59 (4): 330-332.
- Lal K., Purohit A. and Ram H. (2017): Insulin mimetic and pancreas-protective effect of *Tecomella undulata* leaves extract in diabetic rats. *World J of Pharmacy and Pharmaceutical Sciences.* 6(2): 924-938.
- Ley S.H., Hamdy O., Mohan V. and Hu F.B. (2014): Prevention and management of type-2 diabetes: Dietary components and nutritional strategies. *Lancet.* 383 (9933): 1999-2007.

- Melendez-Ramierz L.Y., Richard R.J. and Cefalu W.T. (2010): Complications for type-1 diabetes: Review. *Endocrine Metab Clin North Am.* 39: 625-640.
- Okur M.E., Karantas I.D. and Siafaka P.I. (2017): Diabetes Mellitus: A review on pathophysiology, current status of oral medications and future perspectives. *Acta Pharm. Sci.* 55(1): 61-82.
- Pandit R., Phadke A. and Jagtap A. (2010): Antidiabetic effects of *Ficus religosa* extract in streptozotocin-induced diabetic rats. *J Ethanopharmacol.* 128(2): 462-6.
- Parera P.R.D., Ekanayake S. and Ranaweera K.K.D.S. (2017): Antidiabetic compound in *Syzygium cumini* decocation and ready to serve herbal drink. *Evi. Ba. Coml. & Alter. Med.*.
- Patel D.K., Kumar R., Laloo D. and Hemalatha S. (2012): Natural medicines from plant source used for therapy of diabetes mellitus: an overview of its pharmacological aspects. *As Pac J Trop Dis.* 2: 239-250.
- Pradeep A.R., Garg V., Raju A. and Singh P. (2016): Adjunctive local delivery of *Aloe vera* gel in patients with type-2 diabetes and chronic periodontitis: A randomixed, controlled. *Clinical Trial. J Periodontol.* 87 (3): 268-274.
- Purohit A., Vyas K.B. and Vyas S.K. (2008): Hypoglycemic activity of *Embica ribis* berries (50% EtOH) extract on Alloxan induced diabetic rats. *Ancient Sci of Life.* XXVII (4): 538-542.
- Purohit A. and Sharma A. (2006): Blood glucose lowering potential of *Bougainvillea spectabilis* leaf extract in streptozotocin induced type-1diabetic rats. *Ind Drugs.* 43 (7): 538-542.
- Purohit A. and Daradka H.M.M. (2000): Antidiabetic activity of *Syzygium cumini* seed extract in alloxan induced diabetic rats. *Hamdard Medicus.* XLIII 4: 33-34.
- Samual V.T and Shulman G.I. (2012): Mechanism for insulin resistance: Common threats and missing links. *Cell.* 148: 852-887.
- Shikha B., Prakash N.B. and Nikhil B (2011): Medicinal plants of Rajasthan (India) with antidiabetic potential. *Int Research Journal of Pharmacy.* 2(3): 1-7
- Sharma S. and Agarwal P. (2015): Diabetes mellitus: an ayurvedic view. *J Scient Reas.* 4(4): 193-196.
- Singla S., Kesheri M., Kanchan S. and Aswath S. (2019): Current status and data analysis of diabetes in India. *Int J of Innovat Techno and Exlpori Engine (IJITEE).* 8(9): 1920-1934.
- Solanki M. and Purohit A. (2021): Insulin mimetic and pancreas protective effect of Leptadenia pyrotechnica (forssk) decne. stem in type-2 diabetic albino rats. *Advances in Bioresearch.* 12(3): 192-198.
- Stanley, Mainzen., Prince P., Menon V.P. and Gunasekaran G. (1999): Hypolipidemic action of *Tinospora cardifolia* roots in alloxan diabetic rats. *J of Ethnopharmacology.* 64 (1): 9-15.
- Tridgell D.M., Tridgell A.H. and Hirsh I.B. (2010): Inpatient management of adults and children with type-1 diabetes. Review of ketoacidosis, uncontrolled hyperglycemia and hypoglycemia, including perioperative insulin management. *Endocrin Metab Clin North Am.* 39: 595-608.

- Tziakas D.N., Chalikas G.K. and Kalki J.C. (2005): Epidemiology of the diabetic hearts. *Coron Artery Dis*. 16 (1): S3-S10.
- Wadood A., Wadood N. and Shas S.A. (1989): Effects of Acacia Arabica and Caralluma edulis on blood glucose level on normal and alloxan diabetic rabbits. *J Pakistan Med Assoc*, 39: 208-212.
- World Health Organization (2005): Preventing chronic diseases: A vital investment. WHO global report. Geneva.
- World Health Organisation (1999): Definition, diagnosis and classification of diabetes and classification of diabetes mellitus (Department of Noncomunicable Disease Survellience, Genewa.
- World Health Organization- Global report on diabetes (2010): Geneva.
- Yadav P., Sarkar S. and Bhatnagar D. (1997): Action of Capparis deciduas against alloxan indeced oxidative stress and diabeties in rat's tissue. *Pharmacological Reaearch* 36 (3): 221-228.
- Yadav J.P., Saini S., Kalia A.N. and Dangi A.S. (2008): Hypoglycemic and Hypolipidemic activity of ethanolic extract of *Salvadora oleoides* in normal and alloxan induced rats. *Indian J. Pharmacol*. 40 (1): 23-27.
- Zimmet P. (2000): Globalization, coca-colonization and the chronic disease epidemic: Can the doomsday scenario be averted? *J Int Med*. 247: 301-310.

CHAPTER

13

Antiandrogenic Activities of Medicinal Plants

Lalita Goyal and Ashok Purohit
Department of Zoology, Jai Narain Vyas University, Jodhpur, Rajasthan, India
E-Mail: lalitagoyal025@gmail.com

ABSTRACT

Rapid increase in uncontrolled population distribution has become a big issue for the developing countries like India. The growing population is thereby causing an environment stress by affecting the natural resources, economic and social conditions. Since 19th century the growth rates increased exponentially, reached high peaks in the 20th century. Total world population from 2010 to 2045 is expected to reach 9 billion. Several preventive and effective methods of contraceptive are available for fertility control like hormonal and chemical but they not free from side effects such as obesity, carcinogenic effects, and bloating, and headache. Therefore there is a need to control the human fertility rate by developing in expensive, safe, effective and universally expectable contraceptives. The present study was carried out to find antiandrogenic plant.

Keywords: Population, Fertility, Contraceptive, Antiandrogenic, Preventive.

INTRODUCTION

Rapid increase in uneven population distribution has become a big issue for the developing countries like India. The growing population is thereby causing an environment stress by affecting the natural resources, economic and social conditions. Since nineteenth century the growth rates increased exponentially, reached high peaks in the 20th century. Total world population from 2010 to 2045 is expected to reach 9 billion (Khourdaji *et al.,* 2018). Therefore, there is a need to control the human fertility rate by developing an in expensive, safe, effective and universally acceptable contraceptive.

The Indian Family Welfare Programme is one of the largest public health programmes in the world. India is the first country to have launched a well-defined Family Planning Programme in 1951. The first five-year development plan initiated in 1951 in India recognized the need for population stabilization as an essential prerequisite for sustaining a good quality of life and a National Family Planning Programme was launched in 1952. The main motto of this programme was "reducing birth rate to the extent necessary to stabilize the population at a level consistent with requirement of national economy". The results of the 1961 census, which showed a rapid growth in population, prompted the government to form the Department of Family Planning. The Family Planning Programme was renamed as Family Welfare Programme in 1979 and family planning services were integrated with those of Maternal and Child Health and Nutrition.

Continuous and fast expansion of the Earth's population makes the need for safe and effective measures of contraception. It is generally agreed that contraceptive use is the key to improve reproductive health. Contraceptive prevalence can be improved if a variety of contraceptive technologies are provided to the users backed up by good quality of services. The common goal is to improving the quality of life for all human beings.

Several preventive and effective methods of contraceptives are available for fertility control like hormonal and chemical but they are not free from side effects such as obesity, carcinogenic effects, and bloating, and headache, breast tenderness (Ghogar and Jiraungkoorskul, 2017).

The medical science research has still to continue to develop new safer, reasonable method to control the population explosion. Male Contraception is one of the promising answers to the above problem perhaps. A large number of contraceptive methods are available for women in market but only few methods are available for male which includes condom and vasectomy (Gava *et al.*, 2016). Thus, there is an urgent need to search simple, safe, relatively, non-toxic, reasonable and effective indigenous remedy for male antifertility and that to from the naturally available plant product. Various plants are claimed to possess aphrodisiac potential found in Ayurveda. Vajikarana, is an Ayurvedic branch which deals with infertility problem. Ayurveda has many plant-based medicines to cure the male reproductive dysfunction and Ayurveda also has potential to control fertility via herbal products which act as contraception. Plants and plant-based products have been used as a valuable and safe natural source of medicines for treating various ailments. However, several commonly used plants have been reported to adversely affect male reproductive functions in wildlife and humans. The effects observed with most of the plant and plant-based products have been attributed to the antispermatogenic and antisteroidogenic properties of one or more active ingredients (D'Cruz *et al.,* 2010).

Many number of plant which showing antifertility efficacy have been reported:

***Ocimum sanctum*:** Locally known as tulsi belongs to the famiy Lamiacea. The anti-androgenic property of *Ocimum sanctum* leaves extract at the dose of 250 mg/kg/day for 45 days was studied on albino rat. The extract showed significant decrease in sperm count, motility and increase in sperm anomalies and also showed alteration in the cauda epididymal milieu due to androgen deficiency (Ahmed *et al.*, 2011).

***Azadirachta indica* (L.):** Popularly known as margosa or "neem" belongs to family Meliaceae. Almost all the parts of this tree are been used for medicinal purposes.Neem has long been documented to have antifertility effects (Ogbuewu *et al.,* 2009). Purohit *et al.,* (1990) have reported contraceptive efficacy of neem flower and neem bark in male albino rats. Similarly neem leaves extract also showed contraceptive efficacy in male rats (Purohit, 1999). Administration of the aqueous leaf extract of neem (dose of 200 mg kg−1 for 28 days) damaged the seminiferous tubules, resulting in the slackening of germinal epithelium, marginal scondensation of chromatin in round spermatids, degeneration and learrangement of germ cell types from their orderly arrangement in spermatogenesis in the male mice. The effects were reported to revert back to normal after 42 days of withdrawal of the treatment (Mishra and Singh, 2005).

***Tripterygium* wilfordii:** A Chinese plant which is locally known as thunder duke vine, belonging to the family Celastraceae. It is used for the treatment of psoriasis and rheumatoid arthritis, has been shown to cause necrospermia or azoospermia in patients treated for rheumatoid arthritis. A number of diterpeneepixidesviz, triptolide, tripdiolide, triptodenol, tripochlorolide and 16-hydroxy triptolide have been isolated from *Tripterygium wilfordii.* These compounds act mainly on spermatids and testicular and epididymal spermatozoa with exfoliation and inhibition of basic nuclear protein turnover of late spermatids, delayed spermiation and sperm head-tail separation, microtubule, microfilament and membrane damages. A preliminary toxicity evaluation indicated that these compounds were immunosuppressive at dose level 5-12 times their antifertility doses. Treatment of triptolide at the dose of 100 μg/kg bw/day for 82 days showed a decrease in cauda epididymal sperm content by 84% in male rats (Huyunh *et al.*, 2000).

***Albizza lebbeck*:** The methanolic extract of *Albizza lebbeck* pods at the doses of 50, 100 and 200 mg dry matter/kg.BW were reported to reduce diameter of epididymides (Gupta *et al.*, 2004).

***Piper longum*:** Locally known as long pepper belongs to family Piperaceae. Administration of piperine at the dose 100 mg/kg body weight has been shown a significant reduction in sperm motality, viability and count in male rats (D'Cruz and Mathur, 2005).

Ruta graveolens: Commonly known as sodab belongs to the family Rutaceae. Oral administration of aqueous extract of *Ruta graveolens* with the dose of 500 mg/kg for 60 days has been shown antiandrogenic activity by decrease the weight of genital organ and sperm motality (Khouri and El-Akawi, 2005).

Foeniculum vulgare: Anethole, a major constituent of the plant *Foeniculum vulgare* (Umbellivferae) when administered to male rats at the doses of 70, 140 and 180 mg/kg/day.BW decreased the number of spermatogonia and sperm count and also reduce reproductivity. The compound showed antiandrogenic activity (Mansouri *et al.*, 2016).

***Sarcostemma acidum*:** The methanolic extract of *Sarcostemma acidum* at a dose of 100 mg for 60 days has been shown to cause significant decrease in the number of mature leydig cells and an increase in the degeneration of Leydig cell population in male albino rats (*Verma et al.*, 2002).

Martynia annua: Administration of the ethanolic extract of *Martynia annua* roots at the doses of 100 and 200 mg/kg.BW for 60 days caused Leydig cell atrophy and also showed reduction in the sperm number in male rats (Mali *et al.,* 2002).

Leptadenia hastate: The aqueous extract of *Leptadenia hastate* Decne leaf stem showed significant reduction of the progressive velocity, linearity and sperm motality in rats (Bayala *et al.*, 2011).

Tinospora cordifolia: Administration of 70% methanolic extract of *Tinospora cordifolia* stem at the dose of 100 mg/day for 60 days has been shown the significant reduction in Leydig cell nuclear area and mature leydig cell numbers (Gupta and Sharma, 2003).

Colebrookea oppositifolia: The ethnolic extract of *Colebrookea oppositifolia* at the dose 200 mg/day for 10 weeks were reported to cause a decrease in the nuclear and cytoplasmic surface area of Leydig cells in male rats (Gupta *et al.*, 2001).

***Cissampelas pareira*:** The aqueous extract of *Cissampelas pareira* leaves at the dose of 1 ml/100 g. BW for 14 days in albino rats reduced the seminal spermatozoa and increased the percentage of abnormal spermatozoa (Luangpirom *et al.,* 2010).

Bougainvillea spectabilis: Administration of the aqueous extract of *Bougainvillea spectabilis* leave has been shown significant decrease in the number of spermatozoa and athrophied seminiferous tubules in Swiss albino mice (Mishra *et al.*, 2009).

Melia azadrach : The rats treated with seed oil of *Melia azadrach* at the doses of 50 and 100 mg dry matter/kg. BW for 60 days were showed decrease in live spermatozoa and following 90 days withdrawal period, showed normal number of spermatozoa (Parandin *et al.*, 2008).

Carica papaya: Degeneration of germ cells and germinal epithelium, reduction in the number of Leydig cells and presence of vacuoles in the seminiferous tubules were demonstrated when crude ripe seeds of *Carica papaya* were administered orally to male Wistar rats at a dose of 100 mg per kg body weight (Ogbuewu, *et al.,* 2010).

Abrus precatorious: Popularly known Ratii belongs to family Leguminosae. The aqueous extract of *Abrus precatorious* seed at the doses of 400,800 and 1600 mg/kg.BW for 18 days showed testicular degeneration due to reduction in the number of sperm cells (Adedapo *et al.*, 2007).

Allium sativum: Is commonly known as garlic belonging to the family Amaryllidaceae. The crude aqueous extract of Allium sativum at the doses of 0.25 and 0.50 g/ml have been reported to reduce sperm viability, membrane disintegration of sperm and irreversible immobilization of epididymal and human ejaculated sperm (Chakrabarti *et al.*, 2003). Administration of the crude extracts of garlic for 30 days showed a significant reduction in the level of serum testosterone in male rats (Hammami *et al.*, 2008).

Thevetia peruviana: Belongs to the family Apocynaceae. The active constituents of *T. peruviana* are α-amyrin acetate, lupeol acetate, α-amyrin, β-amyrin, lupeol and thevetigenin. The male rats treated methanolic extract of *T. peruviana* stem bark at the dose of 100 mg/day was showed significant reduction in the weight of reproductive organs and also showed a significant fall in the total protein and sialic acid content of the testes, epididymides, seminal vesicle and ventral prostate, as well as in the glycogen content of testes (Gupta *et al.*, 2011).

Cuminum cyminum: Saxena *et al.,* (2015) reported that oral dose of *Cuminum cyminum* fraction at 50 mg/rat/day for 60 days inhibited spermatogenesis in rats due to decreased counts in round spermatids, preleptotene spermatocytes, secondary spermatocytes, testosterone and number of mature leydig cell were also decreased.

Parkinsonia aculeata (Linn.) belongs to the family Caesalpiniaceae, also knownvilayti babul. The ethanolic extract of *Parkinsonia aculeate* stem bark was administered orally to male rats at the dose of 50, 100 and 200 mg/day for 60 days showed significant suppression of cauda epididymal sperm count and motality and also showed reduction in primary and secondary spermatocytes, serum testosterone level, seminiferous tubular diameter and count of Leydig cells (Gupta *et al.*, 2007).

***Momordica charantia*:** Administration of the ethanolic extract of *Momordica charantia* at the doses 400 and 800 mg/kg.BW/day showed significant reduction in diameters of seminiferous tubules and epididymides, spermatid density, daily sperm production and caudal epididymal spermatozoa, sperm motility and viability (Abalaka *et al.,* 2009).

Citrullus colocynthis: Antifertility property of *Citrullus colocynthis* (indrayan) belongs to family Cucurbitaceae has been claimed in the male rat 50% ethanolic fruit extract caused a significant decrease in sperm density and motility along with marked reduction in sperm cell dynamics. (Chaturvedi *et al.*, 2003).

Butea monosperma: belongs to the family leguminacea. It is commonly known as the flam of the forest or Palas. Active constituents of *Butea monosperma* is butrin, also contains flavonoids and steroids. Methanolic extract of *Butea monosperma* leaf (doses 200 and 400 mg/kg.BW) for 21 days significantly reduced the sperm motility, sperm count and also reduced reproductive organ weights in male albino rats (Reddy and Rajendran, 2018).

Coccinia indica : Ethanolic extract of *Coccinia indica* leaf at the doses of 200 and 500 mg/kg.BW per day for 56 days has been shown adverse effect on serum level of testosterone, steroidogenic markers in the testis and on sperm parameters in the cauda epididymidis (Verma and Singh, 2017).

Table: List of Indian plants reported for their antiandrogenic and antispermatogenic potential

S. No.	Name of the plant	Family	Common name	Plant parts used	Dosage	Model used	Active constituents	Effect
1.	*Tripterygium wilfordii*	Celastraceae	Thunder duke vine	Leaf	100 u/mg/kg bw for 82 days.	Male Wistar Rats.	Triptolide, Tripdiolide, Triptodenol, Tripo-chlorolide.	Antiandro-genic activity via significantly effect on spermatid and testicular and epididymal spermatozoa.
2.	Albizza lebbeck	Fabaceae	Fry wood	Benth pods	Methanolic extract at 50,100 and 200 mg dry matter/kg. bw	Male Wister Rats	Melacacidin, D-catechine, B-sitosterol, Betulnic acid	reduce diameter of epididymides.
3.	*Leptadenia hastate*	Lauraceae	Tejpatta, Indian bay leaf	Leaf	Aqueous extract	Wister rats	Alkaloids, phenolic compounds such as linalool, β Caryoohyllene.	Antispero-matogenic effect (reduction of the progressive velocity, linearity & sperm motality).
4.	*Thevetia peruviana*	Apocynaceae	Yellow oleander	Stem Bark	Methanolic stem extract 100 mg/day	Male albino rats.	A-amyrin acetate, lupeol acetate, A-amyrin, lupeol and thevetigenin	Antispero-matogenic effect (reduction in the weight of reproductive organs).
5.	*Cuminum cyminum*	Apiaceae	Jeera	Seed	50 mg/rat./day for 60 days	Rats	Cumin	Antispero-matogenic effect (decreased counts in round spermatids, preleptotene spermatocytes, secondary spermatocytes & number of mature leyding cell were also decreased).

6.	*Butea monosperma*	Leguminacea	Flam of the forest or palas	Leaf	Methanolic leaf extract 200 and 400mg/kg for 21 days.	Male albino rats	Butrin	Antispero-matogenic effect.(reduced the sperm motility, sperm count & reproductive organs).
7.	*Coccinia indica*	Cucurbitaceae	Lvy gourd Kundru	Leaf	Ethanolic leaf extract 200 & 500 mg/kg for 56 days.	Male albino rats	Alkaloids Flavonoids	Adverse effect on serum level of testosterone, steroidogenic markers in the testis & sperm parameters in the cauda epididymidis.
8	*Azadirachta indica*	Meliaceae	Neem	Bark flower	Aqueous extract of 200 mg/ kg/day 28 days	Male albino mice	Azadirachtin, Nimbin, Nimbidin, and Salannin	Decrease sperm mortality & sperm count. siginificant increase of abnormal sperm.
9.	*Sarcostemma acidum*	Asclepiadaceae	Moon plant, Soma	Flower twig	Methanolic extract at 100 mg for 60 days.	Male albino rats	Lupeol, a- amyrins, B-amyrins.	Significant decrease in the number of mature leydig cells.
10.	*Momordica charantia*	Cucurbitaceae	Karela, Bitter guard	Seed	Ethanolic extract 400 & 800 mg/ kg/day 42 days	Male Wister rats	Charantin, a steroidal glycoside.	Reductions in diameters of seminiferous tubules and epididymides, spermatid density, daily sperm production and caudal epididymal spermatozoa, sperm motility.
11.	*Martynia annua*	Martyniaceae	Iceplant	Root	ethanolic extract 100 and 200 mg/kg/ body wt. for 60 days.	Male albino rats	Linoleic acid, Palmitic acid.	Reduction in the serum concentration of LH and testosterone.
12.	*Tinospora cordifolia*	Myrtaceae	Amrita,giloe	Stem	Methanolic extract 100 mg/kg for 60 days.	Male albino rats	Terpenoid, Alkaloid.	Significant reduction in leydig cell nuclear area and mature leydig cell numbers.

13.	*Allium sativum*	Amaryllidaceae	Garlic	Bulb	Methanol extract 500,1000 mg/kg body weights	Albino male mice	Alliin,Allicin, allylpropl, diallyl, trisulfide, sallylcysteine, vinyldithiines, S-allylmer-captocystein	Antifertility activity
14.	*Ocimum sanctum*	Lamiacea	Tulsi	Leaf	250 mg/kg/ day for 45 days	Albino male rats	Oleanolic acid, rosmarinic acid, ursolic acid, linalool, carvacrol	Decrease in sperm count and motility (antiandrogenic effect)
15.	*Colebrookea oppositifolia*	Lamiaceae	Indian squirrel tail	Leaf	Ethnolic extract 200mg/ day for 10 weeks.	Male rats	Stearic acid, Palmitic acid,oleic acid.	Decrease in the nuclear and cytoplasmic surface area of leydig cells.
16.	*Piper longum*	Piperaceae	Black paper	Fruit	10 mg/kg for 60 days	Albino male rats	Piperine	Effect on epididymal sperm count, motility & reproductive organs.
17.	*Bougain-villea spectabilis*	Nyctaginaceae	Great bougainvillea	Leaf	aqueous extract.	Albino mice	Methyl salicylate	Significant decrease in the number of spermatozoa and athrophied seminiferous tubules.
18.	*Cissampelas pareira*	Menisper-maceae	Velvet leaf	Leaf	1ml /100g BW for 14 days.	Albino rats	Alkaloids, Flavonoids	Reduced the seminal spermatozoa and increased the percentage of abnormal spermatozoa.
19.	*Melia azadrach*	Meliaceae	Bead tree, Bakain , Chinaberry tree.	Seed oil	50 & 100 mg/ dry matter /kg for 60 days.	Male albino mice	Terpenoids, Limonoids.	Decrease in live spermatozoa.
20.	*Abrus precatorious*	Leguminosae	Asghungchi, Ratti.	Seed	Aqueous extract at 400,800 & 1600 mg/ kg for 18 days.	Male albino rats	Luteolin, Abrectorin.	Showing testicular degeneration due to reduction in the number of sperm cells.

21.	*Parkino-sonia aculeate*	Caesalpi-niaceae	Vilayti babul	Stem bark	Ethanolic extract at 50, 100 and 200 mg/ kg body weight for 60 days,	Male rats	Alkaloids, Flavonoids	Significant suppression of cauda epididymal sperm count and motality and also showed reduction in primary and secondary spermatocytes, serum testosterone level.
22.	*Citrullus colocynthis*	Cucurbitaceae	Indrayan	Fruit	Used 50% Ethanolic extract.	Male Rats	Alkaloids, Flavonoids	Significant decrease in sperm density and motility along with marked reduction in sperm cell dynamics.
23.	*Foeniculum vulgare*	Apeaceae	Saunf, fennel	Seed	hydro-alcoholic extract 35, 70, 140, and 280 mg/kg/b.w. daily for 60 days.	Male Wister rats	Anathole, Estragole.	reduce reproductivity and has anti-fertility activity.

CONCULUSION

It can be concluded that plant-based phytomedicines are safe, easily available and cost effective approaches for antifertility activities. Due to harmful and health disasters effects of synthetic drugs, there is need to more research for antifertility action of plant products.

ACKNOWLEDGMENT

We are thankful to the Head and Department of Zoology, Jai Narain Vyas University, Jodhpur, Rajasthan. Heartfelt thanks to Professor Ashok Purohit and Professor Kalpana Purohit for inspiring us to write this paper. Special thanks to Dr. H. Ram Barad and Dr. R.P. Saran.

REFERENCE

- Abalaka, M.E., Olonitola, O.S., Onaolapo, J.A. and Inabo, H.I. (2009): Evaluation of acute toxicity of *Momordica charantia* extract using Wister rats to determine safety levels and usefulness of the plant in ethnochemotherapy. *Int. J. P. App. Sci.*, 3:1–6.
- Adedapo, A.A., Omoloye, O.A. and Ohore, O.G. (2007): Studies on the toxicity of an aqueous extract of the leaves of *Abrus precatorius* in rats. Onderstepoort. *J. Vet. Res.*, 74:31–36.
- Ahmed, M., Ahamed, R.N.H., Aladakatti, R. and Ghodesawar, G.M. A. (2011): Effect of benzene extract of *Ocimum sanctum* leaves on cauda epididymal spermatozoa of rats. *Iranian Journal of Reproductive Medicine*, 9(3):177-86.

- Bayala, B., Telefo, P.B., Bassole, I.H.N., Tamboura, H.H., Belemtougri, R.G., Sawadogo, L., Malpaux, B. and Dacheux, J.L. (2011): Anti-spermatogenic activity *Leptadenia hastate* leaf stems aqueous extracts in male Wister rats. *J. Pharmacol. Toxicol.*, 1–9.
- Chakrabarti, K., Pal, S. and Bhattacharyya, A.K. (2003): Sperm immobilization activity of *Allium sativum* L. and other plant extracts. *Asian J. Androl.*, 5:131–135.
- Chaturvedi, M., Mali, P.C. and Ansari, A.S. (2003): Induction of reversible antifertility with a crude ethanol extract of *Citrullus colocynthis* Schrad fruit in male rats. *Pharmacology*, 68:38–48.
- D'Cruz, S.C. and Mathur, P.P. (2005): Effect of piperine on the epididymis of adult male rats. *Asian J. Androl.*, 7:363–368.
- D'Cruz, S.C., Vaithsinathan, S., Jubendradass, R. and Mathur, P.P. (2010): Effects of plants and plant products on the testis. *Asian Journal of Andrology,* 12: 468–479.
- Gava, G., Lantadilla, C., Martelli, V., Fattorini, A., Seracchioli, R. and Meriggiola, M.C. (2016): Hot issues in female and male hormonal contraception. *Minerva Ginecol* ,68(1):78-89.
- Ghogar, A. and Jiraungkoorskul, W. (2017): Antifertility effect of *Bougainvillea spectabilis* or Paper Flower. *Pharmacognosy reviews*, 11(21):19-22.
- Gupta, R., Kachhawa, J.B., Gupta, R.S., Sharma, A.K., Sharma, M.C. and Dobhal, M.P. (2011): *Hum Fertil. Camb., 14(1):53-9.*
- Gupta, R.S., Rehwani, H., Khushalani, V., Tanwar, K. and Joshi, Y.C. (2007): Antispermatogenic effects of *Parkinsonia aculeata* stembark in male rats. *Pharmaceutical Biology*, 45(1):1-8.
- Gupta, R.S., Kachhawa, J.B. and Chaudhary, R. (2004): Antifertility effects of methanolic pod extract of *Albizzia lebbeck* (L.) Benth in male rats. *Asian J. Androl.*, 6:155–159.
- Gupta, R.S. and Sharma, A. (2003): Antifertility effect of *Tinospora cordifolia* (Wild) stem extract in male rats. *Indian J. Exp. Biol.*, 41:885–889.
- Gupta, R.S., Yadav, R.K., Dixit, V.P. and Dobhal, M.P. (2001): Antifertility studies of *Colebrookia oppositifolia* leaf extract in male rats with special reference to testicular cell population dynamics. *Fitoterapia*, 72:236–245.
- Hammami, I., Nahdi, A., Mauduit, C., Benahmed, M. and Amri, M. (2008): The inhibitory effects on adult male reproductive functions of crude garlic (*Allium sativum*) feeding. *Asian J. Androl.*, 10:593–601.
- Huynh, P.N., Sinha, H.A.P., Wang, C., Stefonovic, K., Leung, A., Atienza, V., Baravarian, S., Reutrakul, V. and Swerdloff, R.S. (2000): Long-term effects of triptolide on spermatogenesis, epididymal sperm function, and fertility in male rats. *Journal of Andrology*, 21(5):689–699.
- Khouri, N.A. and El-Akawi, Z. (2005): Antiandrogenic activity of *Ruta graveolens* L. in male albino rats with emphasis on sexual and aggressive behavior. *Neuro. Endocrinol. Lett.*, 26:823-829.
- Khourdaji, I., Zillioux, J., Eisenfrats, K., Foley, D. and Smith, R. (2018): The future of male contraception: a fertile ground. *Translational Andrology and Urology*, 7(2): S220-S235.

- Luangpirom, A., Sirisarn, W. and Pontaisog, J. (2010): Antifertility activity of the aqueous leaf extract of *Cissampelos pareira* in male Albino mice. *ABAH Bioflux.*, 2:59–64.
- Mali P.C., Ansari A.S. and Chaturvedi M. (2002): Antifertility effect of chronically administered *Martynia annua* root extract on male rats. *J. Ethnopharmacol.*, 82:61–67.
- Mishra, N., Joshi, S., Tandon, V.L. and Munjal, A. (2009): Evaluation of anti-fertility potential of aqueous extract of *Bougainvillea spectabilis* leaves in Swiss Albino mice. *IJPSDR.*, 1:19-23.
- Mishra, R.K. and Singh, S.K. (2005): Effect of aqueous leaf extract of *Azadirachta indica* on the reproductive organs in male mice. *Indian J. Exp. Biol.*, 43:1093–103.
- Mansouri, E., Asadi-Samani, M., Kooti, W., Ghasemiboroon, M., Ashtary-Larky, D., Alamiri, F., Afrisham, R. and Noohi, Z. (2016): Anti-fertility effect of hydro-alcoholic extract of fennel (*Foeniculum vulgare* Mill) seed in male Wister rats. *Journal of Veterinary Research*, 60(3):357-363.
- Ogbuewu, I.P., Okoli, I.C. and Iloeje, M.U. (2009): Semen quality characteristics, reaction time, testis weight and seminiferous tubule diameter of buck rabbits fed neem (*Azadirachta indica* A Juss) leaf meal based diets. *Iran. J. Reproduct. Med.*, 7:23–28.
- Ogbuewu, I.P., Agoh, C., Etuk, I.F., Okoli, I.C. and Iloeje, M.U. (2010): Evaluation of the cholesterol characteristics of the liver of Dutch Belt rabbit bucks fed toasted commercial poultry diets supplemented with soy milk. *Int. J. Agric. Res. Dev.*, 2(1):77–81
- Parandin, R., Sadeghipour, H.R. and Rohani, S.A.H. (2008): Evaluation of antifertility effect and recovery of the seed oil constituents of Iranian species of *Melia azadrach* L. in male rats. *J. Reprod. Contracept.,* 19:161–166.
- Purohit, A., Joshi, V.B. and Dixit, V.P. (1990): Contraceptive efficacy of *Azadirachta indica* (flower and bark) in male rats: A biochemical and sperm dynamics analysis. *J. Curr. Biosci.*, 7(4):129-133.
- Purohit, A. (1999): Contraceptive efficacy of neem leave (*Azadirachta indica* A. Juss) in male rats. *Aryavaidan,* 12:231-233.
- Reddy, A.S. and Rajendran, S. (2018): Effect of methanolic leaf extract of *Butea monosperma* on the fertility of male albino rats. *World Journal of Pharmacy and Pharmaceutical Sciences,* 7(5):1588-1597.
- Saxena, P., Gupta, R. and Gupta, R.S. (2015): Contraceptive studies of isolated fractions of *Cuminum cyminum* in male albino rats. *Natural Product Research* 29:24, 2328-2331.
- Verma, P.K., Sharma, A., Mathur, A., Sharma, P. and Gupta, R.S. (2002): Effect of *Sarcostemma acidum* stem extract on spermatogenesis in male albino rats. *Asian J. Androl.*, 4:43–47.
- Verma, H.P. and Singh, S.K. (2017): Antifertility efficacy of *Coccinia indica* in male mice and its possible mechanisms of spermatogenesis. *General and Comparative Endocrynology*, 241:89-99.

CHAPTER

14

Atherosclerosis, It's Major Risk Factors and Ayurveda: A Long Way to Go

Harshlata Chouhan *, Ashok Purohit
Department of Zoology, Jai Narain Vyas University, Jodhpur, Rajasthan, India.
Email: harshlata.chouhan@gmail.com

ABSTRACT

Atherosclerotic plaque development results in predisposition of cardiovascular disease. These diseases have been the harbinger cause of deaths and morbidity in the world. During recent time when the world is in grief of the pandemic of SARS CoV-2 virus, people who are already dealing with cardiovascular diseases are at major risk for the encounter of the infection as well casualties are more prominent in these patients. The onset of atherosclerotic plaque is believed to be the result of high cholesterol concentration and action of mild inflammatory response in the body. There are several risk factors which help in assuming the future likelihood of manifestation of cardiovascular diseases. These risk factors can be classified as traditional and non-traditional risk factors. Traditional risk factors include age, gender, alcohol intake, insulin resistance, lipid concentration, smoking. While non-traditional biomarkers include plasma homocystein level, C-reactive protein, fibrinogen, oxidative stress, leukocyte count and leptin hormone concentration in body. The allopathic treatment of atherosclerosis vastly comprises statins, niacin, beta blockers and fibrates. Although these allopathic medicines are used widely for the treatment of the disease, but long time consumption of these medicines cause may result in predisposition of several side effects. On other hand there are various plants rich in bioactive compounds that have prominent results on biomarkers of atherosclerosis. Many medicinal plants have anti-inflammatory, hypo-lipidemic and cardio protective properties which can be used as safe, trustworthy and cost effective treatment of disease.

Keywords: Atherosclerosis, Hyperlipidemia, Ayurveda, Markers of atherosclerosis.

INTRODUCTION

Atherosclerosis is a disease which is precisely entering its root in human population day by day. Atherosclerotic diseases are considerably responsible for mortality and morbidity in several couple of decades around the world (Fadini and Avogadro, 2007). The term atherosclerosis signifies a condition in which hardening or narrowing of arteries takes place. The process of narrowing of the arteries occurs as the wall of artery starts developing an abnormality; this abnormal structure is called lesion. The lesion further develops slowly and becomes atherosclerotic plaque. Fat, cholesterol, foam cells, smooth muscle cells, calcium and other blood cells are the major constituents of the plaque. Initially, the plaque remains silent and does not show its ill effects, but in later stage as the plaque grows older and bigger in size several other diseases such as coronary artery disease, stroke, peripheral artery disease or kidney problem start emerging (Tibaut *et al.,* 2019). The severity of the disease grows towards lethality and may result in death of the individual.

Progression

Although the disease is conjectured as the disease of modern lifestyle and modern human era but there are significant traces that manifest the prevalence of the disease in ancient time as well (Strong and McGill, 1962). In a study Czermark and Ruffer found the evidences of atherosclerosis in several autopsies of Egyptian mummies, who lived 1000 years before the Common Era (Thompson, *et al.,* 2013). Even in Indian ancient literature there are evidences of prevalence of various heart diseases including atherosclerosis (Grihpunje. *et al.,* 2018). Thus it can be inferred that atherosclerosis is not the disease of current era. It has been a part of human life for a long time and it's been a long journey of countless scientific studies and discoveries that today we have a lot of detailed evidences and remedies for this disease.

The term 'Atherosclerosis' was introduced by Felix Marchand in 1904. Since then the disease has been a serious concern for the medicinal society. In 1908, it was Ignatowski who predicted the relationship between cholesterol rich food and experimentally induced atherosclerosis. Windaus in 1910 and Anichkov in 1913 suggested that how cholesterol was a solo vicious compound in development of atherosclerotic plaque (Konstantinov, *et al.*, 2006). While for a long period of time studies revolved around the fact that plaque development and atherosclerosis was cholesterol storage disorder (Vastesaeger. And Delcourt, 1962). In the modern era of the cell biology of atherosclerosis whole scenario changed and new discoveries revealed that inflammatory process also has a prominent effect in development and progression of atherosclerotic plaque. Rudolf Virchow was one of the pioneer person who established the fact that atherosclerosis is not a simply cholesterol storage disease but it is a complex inflammatory process (Libby, 2012; Geovanini and Libby, 2018). It is assumed that systemic and local inflammatory events mediate all stages of plaque progression. During the period of plaque development pro-inflammatory cytokines, bioactive lipids, inflammatory signaling pathways, leukocytes and several adhesion molecules play a significant role (Mullenix, *et al.,* 2005). Any endothelial wall injury in arterial wall cause endothelial dysfunction which further initiates a cascade of inflammatory process. As the inflammatory machinery gets switched on lymphocyte and monocytes present in blood flow starts binding to the ruptured endothelium and infiltrate arterial wall. Various other cells and molecules such as cytokines, macrophages, T & B lymphocytes, dendritic cells, endothelial cells, vascular smooth muscle cells, interleukins, cell adhesion molecules and tumor necrosis factors play a crucial role in this process (Zhu *et al.,* 2018). With all these inflammatory ingredients low density lipoproteins and very low density lipoproteins add twist to the whole scenario as at the site of plaque development a large number of LDL particles get oxidized and gets converted into Ox LDL. This Ox LDL particles have higher tendency to penetrate arterial endothelial wall and it tends to accumulate there. The monocytes engulf the Ox LDL rapidly and become foam cells, which act as one of the major constituents of atherosclerotic plaque. As the time passes the plaque becomes older and the burden of accumulated foam cells, lymphocytes and inflammatory components increase significantly (Chistiakov *et al.,* 2018). The inflammatory component of the plaque secretes matrix metalloprotease (MMPs) enzyme. MMPs are the proteinase that degrades collagen fibers of intracellular matrix of plaque. As the normal blood flow in the arteries gets obstructed by the plaque, blood platelets also start aggregating at the site of hindrance. This platelet aggregation activates thrombin formation. This is the most dangerous state of the disease. Thus slowly and progressively atherosclerotic plaque hinders the normal blood flow in the corresponding arteries (Williams and Tabas, 2002). The atherosclerotic plaque remains in quiescent state for a long period of time (sometimes even for decade) until unless it produce major turbulence in normal blood flow or there is plaque rupture due to thrombosis. The whole process starts with a mild unrecognizable fatty streak or lesion, and ends up with a hazardous thrombosis which may result in sudden cardiac arrest, myocardial infarction or sudden death.

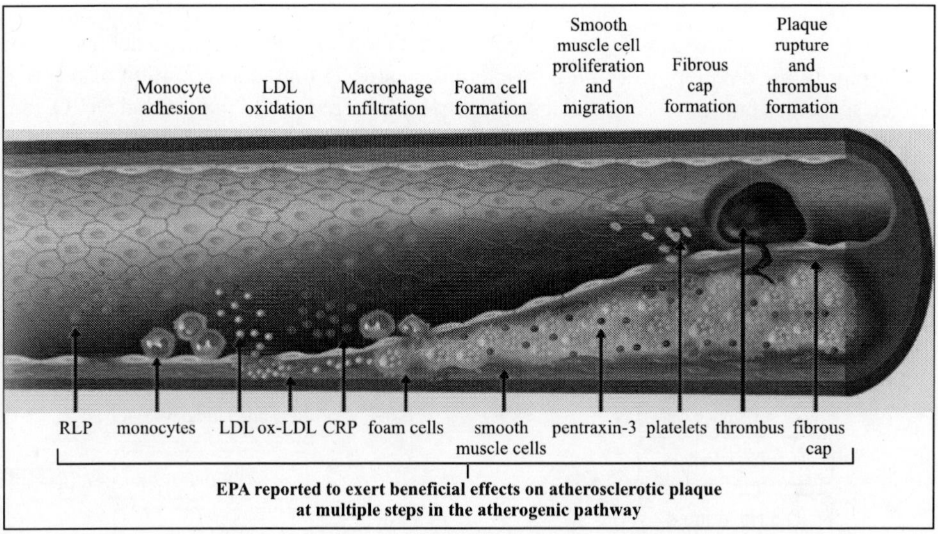

Fig. 1. *Initiation and progression of atherosclerosis(Nelson et al., 2017).*

There are certain markers which may give us major insights about the probability of plaque development in body. As mentioned above, plaque development is a mutual play between elevated lipid concentration in blood and inflammatory machinery of the body. Medicinal fraternity uses certain non-inflammatory and inflammatory markers to predict the upcoming risk of plaque development. These markers are also classified as traditional and non-traditional markers. Here are some of the major markers of atherosclerosis:-

Diabetes/Impaired Insulin Resistance

Diabetes mellitus is considered as a disorder in which carbohydrate metabolism impairment takes place which culminates in an raised level of sugar in the blood or chronic hyperglycemia. This hyperglycemic condition takes place due to improper secretion of insulin hormone or defective insulin action. Diabetes mellitus is an utter risk factor for atherosclerosis occurrence. (Poznyak *et al.,* 2020). A number of studies suggest that a person with diabetes have an elevated burden of subclinical coronary atherosclerosis (Shah *et al.,* 2015).

Lipid Risk Factors

There are substantial evidences that support the relation between elevated lipid levels and threat of atherogenesis in near future. Several parameters of lipid profile have dissimilar impact on body hemostasis. For an instance serum triglyceride concentration alone influence cholesterol metabolism severely. Hypertriglyceridemia act as an independent risk factor for the onset of CVDs. The cause behind this correlation may be due to the fact that triglyceride rich lipoproteins are more prone to metabolic alterations. (Peng *et al.,* 2017). Similarly LDL and VLDL particles are more vulnerable to metabolic alterations and they also exhibit the tendency of passing the endothelial wall barriers (Michos *et al.,* 2019).

Oxidative Stress

Oxidative stress has capability to provoke endothelial dysfunction and act as major contributor to the development of atherosclerotic plaque. A high concentration of reactive oxygen species is produced in vascular endothelial cells. These ROS may include H_2O_2 ion, superoxide anion and various lipid redicals. The predominant cause of generation of ROS in vascular tissue is NAD(P)H oxidase. This increased reactive oxygen species attacks and ruptures endothelial cell lining and thus predispose the machinery for the development of plaque (Ito *et al.*, 2019).

Along with all these traditional risk factors age, alcohol uptake, hypertension, smoking (Fu *et al.,* 2021) and sex are some other traditional risk factors that gives us inside about the internal milieu of body and affects progression of atherosclerosis (Ding *et al.,* 2019).

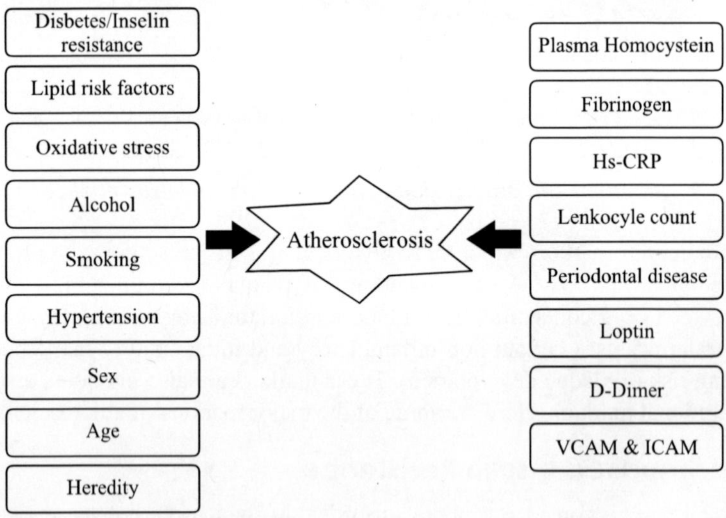

Atherosclerosis: It's major risk factors. Traditional and non-traditional risk factors of atheresc lerso.

Plasma Homocystein

Hyperhomocystenemia or elevated level of plasma homocysteine has been considered as a risk factor. Elevated concentration of homosysteine disturbs HDL-Cholesterol and triglyceride metabolism in the body. Several studies observed that homocysteine may reduce the level of HDL-C in blood via inhibiting the apoA- I protein (Ji *et al.*, 2019). Hcy also accelerates lipid peroxidation and also helps in producing reactive oxygen species. In this way homocysteine directly initiates endothelial dysfunction (Duygu, 2017). The story does not end up here as hcy also gets involved in various other biological mechanisms which directly or indirectly supports plaque formation and generate condition of atherosclerosis (Kubota *et al.*, 2019).

Fibrinogen

As fibrinogen has close relationship with inflammatory cascade it is observed that there is positive relation between fibrinogen activity and plaque development. Fibrinogen is a glycoprotein which is actively involved in the process of thrombosis at the time of injury (Cerit, 2017). Atherosclerosis is also a result of inflammatory cascade, increased level of fibrinogen in serum directly affects the plaque formation and thrombosis of the plaque, making it more hazardous. Fibrinogen tend to increase blood viscosity, stimulates fibrin activation and platelet aggregation at the site of plaque formation (Kryczka *et al.*, 2021).

Hs CRP

High sensitivity C-reactive protein (HsCRP) plays an important role during inflammatory processes of body. In general the level of C-reactive protein increases in case of severe inflammatory conditions. In case of atherosclerosis mild and long term inflammatory machinery acts for a long period of time. So a very low or moderate level of C-reactive protein should be detected precisely. Even low to moderate level of C-reactive protein indicates a sustained condition of low inflammation in the body and thus shows the likelihood of incidence of the disease (Le *et al.*, 2017).

Leukocyte Count

Total leukocyte count gives an overall idea of internal milieu of the body. Number of total leukocyte tends to increase in inflammatory conditions. Research reveals that increased leucocyte count precisely neutrophils and monocytes count may denote increased inflammation in systemic blood flow. This condition acts as a fore runner mechanism of plaque formation. Leukocyte count can be used as reliable morphological indicator for prolonged inflammatory diseases such as cardiovascular diseases and cancer (Chmielewski and Strzelec, 2018).

Leptin

Leptin is a peptide hormone which is secreted by adipocytes cells of adipose tissue. Leptin is a pleiotropic hormone. It acts apon hypothalamus of pituitary gland and regulates the urge of hunger. Generally in normal weighting person the level of leptin hormone remains in physiologically normal concentration. But in obsessed individuals the concentration of leptin increase drastically as there are more number of adipocyte cells present in the body. This over secretion of the hormone creates a condition of leptin insensitivity. This increased level of leptin hormone have multiple effects on body which include activation of various pro- and anti-inflammatory cytokines, it increases the level of CRPs in blood, increases oxidative stress by generating reactive oxygen species and also elevates blood pressure (Dubey and Hesong, 2006).

D-Dimer

D-dimer is end product of fibrin degradation. As fibrin is the last active compound which is responsible for clotting of blood. There is direct relation between d-dimer concentration and risk of ischemic cardiac attacks. The D-dimer test is an important clinical test which is used to detect acute venous thrombosis. Thus this test indicates the future chances of cardiovascular disease occurrence (Folsom *et al.,* 2016).

VCAM-1 and ICAM-1

VCAM-1 (vascular cell adhesion molecue-1) is an adhesion molecule, which is presented by activated endothelial cells. While ICAM (intercellular adhesion molecule-1) is a glycoprotein which is also known as CD-54 molecule. Both VCAM-1 and ICAM-1 molecule belongs to immunoglobulin superfamily. Both the molecules participate in leukocyte adhesion to the endothelium wall. Thus both molecules play a harbinger role in plaque formation (Fotis *et al.,* 2012).

Thus these are the major traditional and non-traditional risk factors of atherosclerosis; these non-traditional risk factors give a close and more accurate assessment of the disease prevention. Studies suggest that non-traditional risk factors are significantly related with the severity of the disease.

Atherosclerosis, its Medication and Plants

Atherosclerosis is interplay between inflammatory response and metabolic disorder. The major concern about the treatment of the disease revolves around the maintaining cholesterol level up to the desirable concentration in the body. As high cholesterol level is considered as the main cause of the disease maintaining cholesterol level is the main task for the treatment of the disease. The major treatment therapy for the cure of the disease involves inhibitors of HMGCo A enzyme. The enzyme is the central component in the cholesterol biosynthesis pathway. Inhibiting this enzyme activity directly lower downs the cholesterol concentration in the body. Statins (HMG CoA reductase inhibitors) are widely used drugs for the treatment of atherosclerosis. Apart from statins, fibrates, niacin and beta blockers also included in treatment therapy, as they tend to reduce the cholesterol by various other mechanisms. Though all these medications are very effective. But long term consumption of these allopathic medications directly affects other metabolic and physiological activities of the body and thus shows various side effects.

So it is important to look for safe and reliable alternative medications which can be used for long time without any side effect. Various studies have been performed and even in recent time also various studies are going on for the alternative medication therapy for the cure of cardiovascular diseases. Plant and their bioactive compounds are the best candidates for the alternative medicine therapy. Plants have been used from ancient times for the cure of various ailments. In different parts of the globe different therapies were the part of civilization but one thing was common that in all these civilizations the plants were used for the treatment of various diseases widely, reliably and effectively.

Root bark extract of English walnut tree (*Juglans regia*) shows positive effect on platelet aggregation, it shows inhibitory effect on abnormal thrombosis and platelet aggregation (Amirou *et al.,* 2018). In another study conducted by Ram & Purohit seed extract of *Acacia senegal* (Kumat) shows prominent effect on various lipid parameters in experimentally induced hyperlipidemic rabbits (Ram *et al.,* 2014). Cumin seed's ethanol extract also show prominent results on atherosclerosis plaque development and various toxicity parameters (Chouhan and Purohit, 2018). Wheat whole grain extract had cardio-protective and lipid lowering effect, it also showed normal toxicity profile on experimentally induced hyperlipidemia in rabbits (Chadha and Purohit, 2018). In another research rambutan (lychee) peel extract was found to be effective to upsurge leptin efficacy and also lower down the lipid peroxidation course, in this way it showed prominent anti-atherosclerotic efficacy (Rahayu *et al.,* 2018). *Tribulus terrestris* a weedy annual plant which is commonly known as bur, have various saponins in its aqueous extract. Due to the presence of these saponins the extract showed promising results on inflammatory markers as it reduced the activity of VCAM-1, ICAM-1 and P. selectin. In this way the plant extract showed anti-inflammatory properties (Fard *et al.,* 2020). Various studies which concerned other inflammatory markers also suggested prominent effectiveness of pant bioactive compounds such one study evaluated anti-inflammatory effect of microencapsulated garlic and tomato mixture. The mixture of microencapsulated garlic and tomato reduced the D-dimer formation as well as also observed to reduce the activity of other inflammatory parameters (Kim *et al.,* 2018). Hibiscus is a plant enriched with various phytocompounds. The calyx powder of hibiscus was found to be effective on elevated level of Hs CRP. So it can be assumed that the plant has some sort of anti-inflammatory efficacy (Edigheh *et al.,* 2016). The plant bioactive compounds showed positive effect in case of diabetes mellitus also a study performed by Lal and Purohit *Tecomella undulate* showed pancreas protective effect and had hypoglycemic effect in streptozotocin-induced diabetic rats (Lal *et al.,* 2017). *Leptadenia pyrotechnica* (Forssk) had insulin like properties and also showed pancreas protective effect on hyperglycemic experimental rats (Solanki & Purohit, 2021). Other studies performed on various plant extracts demonstrated promising results against hyperlipidemia and cardiovascular diseases ((Dixit *et al.,* 1992; Purohit and Daradka, 1999a; Purohit, 1999; Purohit *et al*., 1999; Purohit and Vyas, 2006). Traditional medicinal importance of plant bioactive compounds have innumerable number of utilities that they are used in numerous physiological disorders. Not just in case of cardiovascular diseases but in other ailments such as hypothyroid and antifertility also plant extracts have shown positive outcomes in experimentally induced animal models. *Withania somnifera* leaf extract had positive effect on hypothyroidism in experimentally induced rats (Purohit & Purohit, 2018). *Acacia senegal* bark extract also had regulative effect on hypothyroidism (Jangid and Purohit, 2019). Invarious plants extract were also used to treat the problem of infertility.

In-fact in this time of pandemic, novel covid-19 virus has created havoc in human population. The world is suffering from this novel virus for last two years and plant products also gave promising results against this Sars Cov-2 infection. Various medicinal plants were recommended by medicinal fraternity to cure this novel virus disease such one study showed that *Withania somnifera* which is rich in bioactive component like somniferine and withanoside had capability to inhibit the virus peptide receptor. Molecular docking study performed on phytocompounds of *Tinosporia cardifolia* (Giloy) and *Ocimum sanctum* (Tulsi) showed promising results in the treatment of covid virus. The ADMET profile prediction also revealed that consumption of these plants was safe and reliable (Shree *et al.,* 2020). Thus it can be clearly inferred that plant products are virtue given by Mother Nature to human beings.

CONCLUSION

Atherosclerosis and related cardiovascular diseases have remained as major concern of medicinal fraternity. The fact is that the atherosclerosis has been a part of human life since ancient times. Cardiovascular diseases are always considered as one of the fore runner cause of deaths around the world. Various risk factors of the disease are considered as the markers of future risk of the predisposition of the disease. Plants are rich source of bioactive compound and they show prominent effect on various biomarkers of atherosclerosis. Therefore our natural wealth of plants products can give us safe, reliable and cost effective medication.

ACKNOWLEDGEMENT

We would like to acknowledge Head, Department of Zoology, Jai Narain Vyas University, Jodhpur, Rajasthan, for providing all the necessary. Jodhpur (Rajasthan) for providing all the necessary amenities.

REFERENCES

- Amirou, A., Bnouham, A., Legssyer, A., Ziyyat, M., Aziz, M., Berrabah, M. and Mekhfi, H. (2018): *Juglans regia* root bark extract on platelet aggregation, bleeding time and plasmatic coagulation: in vitro and ex vivo experiments. *Evidence-Based Complementary and Alternative Medicine,* 2018: 1-7.
- Cerit, L. (2017): Fibrinogen and atherosclerosis. *Arquivos Brasileriosd Cardiologia,* 108: 189-190.
- Chadha, R. and Purohit, A. (2018): Lipid lowering and antioxidant effect of wheat wholegrain extract in hypercholesterolemic rabbits. *Europian Journal of Biomedical and Pharmaceutical Sciences,* 5(1): 541-546.
- Chistiakov, D.A., Melnichenko, A.A., Grechko, A.V., Myasoedova, V.A., and Orekhov, A.N. (2018): Potential of anti inflammatory agents for treatment of atherosclerosis. *Exp Mol Pathol.,* 104: 114-24.
- Chmielewski, P.P. and Strzelec, B. (2018): Elevated leukocyte count as aharbinger of systemic inflammation, disease progression and poor prognosis: A review. *Folia Morphologica*, 77(2): 171-178.
- Chohan, H. and Purohit, A. (2018): Protective effect of Cumin (*Cuminum cyminum*) seed extract on cardiovascular system, toxicity and hematology on hyperlipidemic rabbits: an experimental study. *Asian Journal of Pharmaceutical Sciences,* 11(10): 353-356.
- Ding, N., Sang, Y., Chen, J., Ballew, S.H., Kalbaugh, C.A., Salameh, M.J., Blaha, M.J., Allison, M., Heiss, G., Selvin, E., Coresh, J. and Matsushita, K. (2019): Cigarette smoking, smoking cessation and long term risk of 3 major atherosclerotic diseases. *Journal of the American College of Cardiology,* 74(4):498-507.
- Man, J.J., Beckman, J.A. and Jaffe, I.Z. (2020): Sex as abiological variable in atherosclerosis. *Circulation Research,* 126(9): 1297-1319.
- Dixit, V.P., Jain, P. and Purohit, A.K. (1992): Medicinal uses of Neem (*Azadirachta indica*) in fertility regulation, Diabetes and Atherosclerosis. *Recent Advances* in *Medicinal, Aromatic and Spice Crops;* (2): 463-471.

- Dubey, L. and Hesong, Z. (2006): Role of leptins in atherogenesis. *Exp. Clin. Cardio.,* 11(4): 269-275.
- Duygu, H. (2017): I sthere any link between homocystein and atherosclerosis?. *Journal of Geriatic Cardiology: JGC,* 14(3): 222-226.
- Edigheh, A., Soltani, R., Zolghadr, M., Keshvariand, N. and Sarrafzadegan, N. (2016): Evaluation of the effects of roselle (*Hibiscus sabdariffa* L.) on oxidative stress and serum levels of lipids, insulin and Hs-CRP in adult patients with metabolic syndrome: a double blind placebo-controlled clinical trial. *Complement Integr.,* 13(2):175-180.
- Fadini,G.P. and Avogadro, A., (2007): Endothelial progenitor cells in the natural history of atherosclerotic. *Atherosclerosis*: 194(1):46-54.
- Fard, A.E., Fereydouni, Z., Mansouri, K. and Mostafaie, A. (2020): Effect of *Tibulus terrestris* L. on expression of ICAM-1,VCAM-1, E-Selectin and proteome profile of human endothelial cell in vitro. *Iranian Journal of Immunology.* 17(1): 64-74.
- Folsom, A.R., Gottesman, R.F., Appiah, D., Shahar, E. and Mosely, T. (2016): Plasma D-dimer and incident ischemic stroke and coronary heart disaease. *Stroke,* 47:18-23.
- Fotis, L., Agrogiannis, G., Vlachos, I.S., Pantopoulou, A., Mrgoni, A., Kostaki, M., Verikokos, C., Tzivras, D., Mikhailidis, D.P. and Perrea, D. (2012): Intracellular adhesion molecule (ICAM-1) and Vscular cell adhesion molecule (VCAM-1) at the early stages of atherosclerosis in a rat model. *In vivo,* 26(2): 243-250.
- Fu, X., Zong, T., Yang, P., Li, L., Wang, S., Wang, Z., Li, M., Li, X., Zou, Y., Zhang, Y., Aung, L.H., Yang, Y. and Yu, J. (2021): Nicotine: Regulatory roles and mechanism in atherosclerosis progression. *Food and Chemical Toxicology,* 112154.
- Geovanini, G.R. and Libby, P. (2018): Atherosclerosis and inflammation: Overview and update. *Clin Sci.,* 132(12):1243-52.
- Grihpinje, K.S., Gupta, V. and Singh, O.P., (2018): An ayurvedic aspect of atherosclerosis: A life style disorder., *Inter. J. of Res. Med. Scie. Tech.*
- Ito, F., Sono, Y. and Ito, T. (2019): Measurement and clinical significance of lipid peroxidation as a biomarker of oxidative stress: oxidative stress in diabetes, atherosclerosis and chronic inflammation. *Antioxidant.,* 8(3): 72.
- Jangid, D. and Purohit, A. (2019): *Acacia senegal* bark extract in regulation of hypothyroidism: An experimental approach. *Asian Journal of Pharmaceutical and Clinical Research,* 12(1): 117-120.
- Ji, X., Leng, X., Dong, Y., Ma, Y., Xu, W., Cao, X.P., Hou, X.H., Dong, Q., Tan, L. and Yu, J. (2019): Modifiable risk factors for carotid atherosclerosis: a meta analysis and systematic review. *Annals of Traslational Medicine,* 7(22): 122-128.
- Kim, Y., Kim, K.J., Park, S.Y., Lim, Y., Kwon, O., Lee, J.H. and Kim, J.Y. (2018): Differential responses of endothelial integrity upon the intake of microencapsulated garlic, tomato extract or a mixture: a single-intake, randomized, double-blind, placebo-controlled crossover trial. *Food Funct.,* 9: 5426-5435.
- Konstantinov, I.E., Mejevoi, N. and Anichkov, N.A. (2006): Nikolai N. Anichkov and his theory of atherosclerosis., *Tex. Heart Inst. J.,* 33(4): 417-23.
- Kryczka, K., Kruk, M., Demkow, M., Lubiszewska, B. (2021): Fibrinogen and a triad of thrombosis, inflammation and thr rennin angiotensin system in premature coronary

artery disease in women: A new insight into sex related differences. *Biomolecules,* 11(7): 1036.

- Kubota, Y., Alonso, A., Heckbert, S.R., Norby, F. and Felson, A.R. (2019): Homocystein and incident atrial fibrillation: The atherosclerosis risk in communities and the multi-ethnic study of atherosclerosis. *Heart, Lung and Circulation,* 28(4): 615-622.
- Lal, K., Purohit, A. and Ram, H. (2017): Insulin mimetic and pancreas-protective effect of *Tecomella undulata* leaves extract in diabetic rats. *World Journal of Pharmacy and Pharmaceutical Sciences*, 6(2): 924-938.
- Le, Y., Zhong, X., Cheng, G., Zhao, C., Zhang, L., Hong, Y., Wan, Q., He, R. and Wang, Z. (2017): HsCRP and all - cause cardiovascular and cancer mortality risk: A meta analysis. *Atherosclerosis,* 259-75-82.
- Libby, P. (2012): History of discovery: Inflammation in atherosclerosis, *Arterioscler Thromb Vasc. Biol.,* 32(9): 2045-51.
- Michos, E.D., McEvoy, J.W. and Blumenthal, R.S. (2019): Lipid management for the prevention of atherosclerotic cardiovascular disease. *New England Journal of Medicine,* 381(16): 1557-1567.
- Mullenix, P.S., Anderson, C.A. and Benjamin, W.S. (2005): Atherosclerosis as inflammation. *Annals of Vascular Surgery.,* 19(1):130-8.
- Nelson, J.R., Wani, O., May, H.T. and Budoff, M. (2017): Potential benefits of eicosapentaenoic acid on atherosclerotic plaques. *Vascular Pharmacology*, 91(2017): 1-9.
- Peng, J., Luo, F., Ruan, G., Peng R. and Li, X. (2017): Hypertriglyceridemia and atherosclerosis. *Lipids in Health and Disease.,* 16(1): 12.
- Poznyak, A., Grechko. A.V., Poggio, P., Myasoedova, V.A., Alfieri, V. and Orekhov, A.N. (2020): The diabetes mellitus-atherosclerosis connection: The role of lipid and glucose metabolism and chronic inflammation. *International Journal of Molecular Sciences*., 21(5): 1835-40.
- Purohit, A. and Daradka, H.M.M. (1999): Hypolipidaemic effect of Neem flower (*Azadirachta indica* A.Juss) in rabbits. *GEOBIOS*; 26(2-3): 146-148.
- Purohit, A. and Daradka, H.M.M. (1999b): Hypolipidemic effect of *Curcuma longa* (Haldi) in rabbits. *Hamdard Medicus*, XLII(3): 26-30.
- Purohit, A. and Purohit, A. (2018): Elroxin like mimic action of *Withania somnifera* leaf extract in hypothyroid-induced rats. *Asian Journal of Pharmaceutical and Clinical Research,* 11(11):280-284.
- Purohit, A. and Vyas, K.B. (2005): Hypolipidaemic efficacy of *Capparis decidua* fruit and shoot extract in cholesterol fed rabbits. *Indian Journal of Experimental Biology.,* 43: 863-866.
- Purohit, A. and Vyas, K.B. (2006b): Antiatherosclerotic efficacies of *Capparis decidua* shoot extract (50% EtOH) in hyperlipidemic rabbits. *Indian Drugs* 43(10): 807-812.
- Rahayu, L.S., Umei, L. and Fitri, A. (2018): Rambutaan peel extract increase leptin efficiency and decrease lipid peroxidation in a rat model of obesity. *Medicinal Plants: International Journal of Phytomedicine and Related Industries,* 10(1): 23-28.

- Ram, H., Jatwa, R. and Purohit, A. (2014): Antiatherosclerotic and cardioprotective potential of *Acacia senegal* seeds in diet induced atherosclerosis in rabbits. *Biochem. Res. Int.,* 2014: 436848.
- Shah, A.D., Langenberg, C., Rapsomaniki, E., Denaxas, S., Rodriguez, M.P., Gale, C.P., Smeeth, L., Deanfield, J., Timmis, A. and Hemigway, H. (2015): Type-2 diabetes and incidence of cardiovascular diseases: A cohort study in 1.90milion people. *The Lancet Diabetes and Endocrinology.,* 3(2): 105-113.
- Shree, P., Mishra, P., Selvaraj, C., Singh, S.K. and Chaube, R. (2020): Targeting COVID-19 (SARS-CoV-2) main protease through active phytochemicals of ayurvedic medicinal plants- *Withania somnifera* (Ahwagandha), *Tinospora cordifolia* (Giloy) and *Ocimum sanctum* (Tulsi)- a molecular docking study. *Journal of Biomolecular Structure and Dynamics,*
- Solanki, M. and Purohit A. (2021): Insulin mimetic and pancreas protective effect of *Leptadenia pyrotechnica* (forssk) decne. stem extract in type-2 diabetic albino rats. *Advances in Bioresearch,* 12(3):181-187.
- Strong, J.P. and McGill, H.C., (1962): The natural history of coronary atherosclerosis., *The American Journal of Pathology,* 40(1):37.
- Thompson, R.C., Allam, A.H., Lombardi, G.P. (2013) Atherosclerosis across 400 years of human history: The horus study of four ancient populations. *The lancet,* 381(9873): 1211-22.
- Tibaut, M., Carpnda, M., Kubatka, P., Sinkovic, A., Valentova, V., Filopova, S., Gazdikova, K. and Gasper, L. (2019): Markers of atherosclerosis: Part:2 – Genetic and imaging markers. *Heart, Lung and Circulation,* 28(5): 578-689.
- Vastesaeger, M.M. and Delcourt, R. 1962: The natural history of atherosclerosis. *Circulation* 26:841-55.
- Williams, K.J. and Tabas, I., (2002): Atherosclerosis and inflammation. *Science.,* 297(5581):521-23.
- Zhu, Y., Xian, X., Wang, Z., Bi, Y., Chen, Q., Han, X., Tang, D. and Chen, R. (2018): Research progress on the relationship between atherosclerosis and inflammation. *Biomolecules,* 8(3):80.

CHAPTER 15

Use of NSAIDS (Non-Steroidal Anti-Inflammatory Drugs) and Threats to Vulture Population

R.P. Saran, Nisha Patel and Manish Joshi
Department of Zoology, Jai NarainVyas University, Jodhpur (Rajasthan), India-342001
Email: saranrp@live.com

ABSTRACT

Vultures are the scavenging birds. The population of vultures is declining due to various causes but the veterinary use of NSAIDS (non-steroidal anti-inflammatory drugs) have adverse impact on scavenging birds. Use of diclofenac, carprofen, flunixin, ibuprofen, phenylbutazone and other NSAIDs are mainly associated with mortality and cause of death of vulture species. Several studies reported negative impact of NSAIDs on other organisms like reptiles, birds and raptors. Some other research studies reported no mortality for the NSAID, meloxicam. Various studies supports the relatively safe use of meloxicam and studies suggested the safer use of this NSAID to replace the most toxic NSAID, Diclofenac , mainly in Asian continent and all over world.

Key words: Vulture, NSAIDs, Diclofenac, Meloxicam.

INTRODUCTION

Background of vultures

Vultures are the scavenging birds which feed on the carcasses of large animals. (Ali and Ripley 1998).Vulture in the country has reduced from 40 million (in 1990) to less than 60 thousand (in 2012). (Prakash, 1999, Gilbert *et al*. 2002 and Prakash *et al*., 2003). Till mid of 1980s vulture were found in large number in India. However, today it is rare to sight a vulture. Vultures are the natural cleaners of the environment. By disposing the dead bodies they check the spread of infectious disease. In absence of vultures, the population of animals, like rodents and stray dogs tends to increase leading to the spread of rabies. Hence, the fast disappearing population of vulture is a serious problem in India and there is a need to protect the vulture for its survival.

About Vulture Species In India

Vultures can take flight to the height approx 6000 to 7000 feet and they can cover more than 100 km distance in a single run. (Bird Life International 2010). There are several species of vulture worldwide out of which 9 species are found in India (Ansari 2015). Species of vultures found in India are- Indian vulture *Gyps indicus* (critically endangered), Indian white rumped vulture *Gyps bengalensis* (critically endangered), Red headed vulture *Sarcogyps calvus* (critically endangered), Selender billed vulture *Gyps tenuirostris* (critically endangered), Egyptian vulture *Neophron percnopterus* (endangered), Cinerous vulture *Aegypius monachus* (near threatened), Bearded vulture *Gypaetus barbatus* (least concern), Griffon vulture *Gyps fulvus* (least concern),and Himalayan vulture *Gyps himalayansis* (least concern). (IUCN-2004, IUCN-2020, Purohit *et al.*, 2013).

About decline of vulture population

There are several cause of the decline of vulture population some of them are habitat destruction, pesticide pollution, slow breading rate, feed in on poisoned carcasses, poisoning of diclofenac and lack of legal protection etc. Among all these reasons to decline, Diclofenac poisoning is the main cause of their huge destruction.

Conservation of Vultures (IUCN)

Replacing diclofenac: There is need to evolve an effective substitute of diclofenac, and the present available substitute meloxicam needs to be subsidized.

Captive-breeding programme: This with aim to reintroduce Vultures into the wild need to be launched on large scale, particularly for Critically Endangered and Endangered species of Vultures.

Legal protection: All efforts should be made to protect and conserve the Near Threatened and Least Concern species of Vultures in India and all the species of Vultures should be legally protected.

In situ conservation: There is need to set up Vulture feeding stations through provision of poison-free food, clean water, bone chips and perches within an open-roofed wire-mesh enclosure for safety and freedom of Vultures.

Habitat restoration: Degraded habitats of Vultures need to be restored.

Protection: Full protection must be given to nests of the Vultures in their breeding habitat.

Toxicity of NSAIDs and vulture population

So many studies or research work done to know the cause of vulture population decline drastically. Studies showed the extent to which toxicity of drugs affects. Critically endangered species of vulture mainly the *Gyps* species has a sudden and huge decline due to NSAIDs intake via dead animal carcasses.

The class of NSAIDS includes:

1. Nonselective COX inhibitors

   Salicylates

   Acetylsalicylic acid (Aspirin)

   Salicylamide

   Pyrazolone derivatives

   Phenylbutazone

   Metamizol (Analginum)

   Indole derivatives

   Indomethacin

   Propionic acid derivatives

   Naproxen

Antranilic acid derivatives

Mephenamic acid

Aryl – acetic acid derivatives

Diclophenac sodium

Oxicam derivatives

Piroxicam

Dihydropyrrolizine carboxylic acid derivative

Ketorolac

2. **Selective COX inhibitors**

Preferential COX-2 inhibitors

Nimesulide

Meloxicam

Nabumeton

Selective COX-2 inhibitors

Celecoxib

Parecoxib

Rofecoxib

Diclofenac is the main NSAIDs which affected the vulture and other scavenging bird population. Use of diclofenac banned in India, Nepal, Pakistan, and other south Asian countries for veterinary purpose in the year 2006. (Swan *et al.,* 2006,Choudhary *et al.*, 2012, Galligan *et al.,* 2020)

Residue of diclofenac persists in the dead animal's body (Cutberth *et al.,* 2011, 2014a, 2014b), vulture scavenges on it and hence the poisonous NSAIDs enter the body of vulture and other scavenging birds. Diclofenac directly affects the kidney and liver of the bird (Saran *et al.*, 2016).

The NSAIDs also use to treat the livestock, like meloxicam as it is meant safer than other NSAIDs (Swann *et al.,* 2006b).The meloxicam is used 90% as Nsaids in Nepal. Toxicity of meloxicam is approx 30% and the toxicity of another Nsaids ketoprofen is approx 70% found in the study in the year 2018. (Naidu *et al.*, 2010,2018).While the drug ketoprofen is already banned in the year 2016 in several countries. (Galligan *et al.,* 2020).

Other NSAIDS like acelofenac, nimesulids, and ibuprofen also considered toxic to the vulture species hence these are also banned partially or completely in the several countries for veterinary purpose. (Cuthberth *et al.*, 2006, Galligan *et al.*, 2020).

Mechanism action of NSAIDs

Mechanism of action Non-steroidal Anti-inflammatory Drugs: NSAIDs control pain by inhibiting cyclooxygenase (COX) enzymes which prevents production of prostaglandins (important local mediators of inflammation that lower the threshold of thermal, mechanical, and chemical nociceptors). However, NSAIDs are also capable of producing gastrointestinal ulceration and bleeding because of inhibition of prostaglandin synthesis. NSAIDs should not be used if there is an indication of renal impairment, hepatic dysfunction, severe hypovolemia, or if gastric ulceration is present.

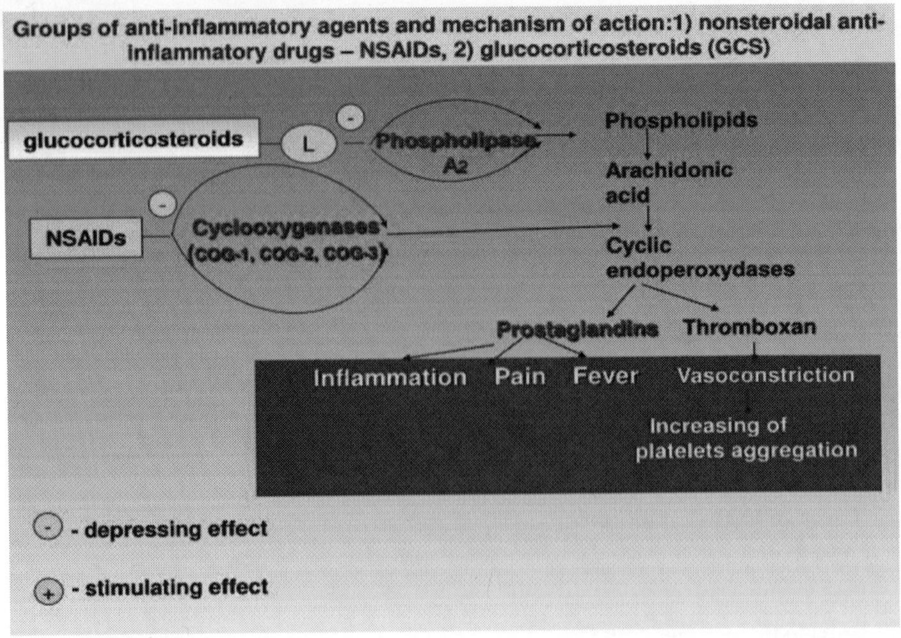

Uses of NSAIDs

NSAIDs provide great benefit in terms of their anti-inflammatory, analgesic, and antipyretic properties; they have also been associated with various adverse side effects. There is an increased risk of gastrointestinal complications ranging from stomach pain to ulcers, risk of heart failure, erectile dysfunction, inflammatory bowel disease and high incidence of renal adverse drug reactions (ADRs). NSAIDs may also cause more severe renal conditions e.g. Interstitial nephritis, Nephritic syndrome, acute renal failure, acute tubular necrosis, may affect the liver or kidneys, photosensitivity reactions, Asthma, Anaemia, Hypertension and regular use may cause hearing loss. If used during pregnancy, it may cause premature closure of the foetal ductus arteriosus and renal ADRs in the foetus and increased risk of premature birth or miscarriage.

Diclofenac sodium is a non-steroidal anti-inflammatory, analgesic and antimicrobial agent and prostaglandin synthesis inhibitor, commonly used in the treatment of rheumatoid arthritis and certain non-rheumatic conditions. Diclofenac can be considered as one of the few NSAIDs of 'first choice' in the treatment of acute and chronic painful and inflammatory conditions.

Clinical uses of NSAIDs

1. Pain: headache, toothache, back pain
2. Fever
3. Arthritis: rheumatoid arthritis, osteoarthritis, gout, ankylosing spondylitis
4. Dysmenorrheal (especially ibuprofen)
5. Unclosure of ductus arteriosus (especially aspirin)

6. Prevention of MI (macular edema deposit of protein and fluid under macula of eye thus eyesight week), stroke, and reinfarction (aspirin)

Pharmacological effects of NSAIDs

- Anti-inflammatory
- Analgesic
- Antipyretic
- Antiplatelet (Aspirin)

Side Effects of Nsaids

Gastrointestinal tract: peptic ulcer and multiple micro-erosions, esophagitis and strinctures, erosive damaging of large and small intestine.

Kidney: reversible acute kidney insufficiency, water electrolyte disorder, chronic kidney insufficiency and interstitial fibrosis, interstitial nephritis, nephritic syndrome.

Cardio vascular syndrome: increasing of arterial hypertension, increasing of static cardiac insufficiency, increase of stenocardia.

Liver: increase of transaminase level, life threatening liver insufficiency.

CNS: headache, somnolence, confusion, disorder of behavior, aseptic meningitis.

Blood system: thrombocytopenia, haemolytic anemia, granulocytopenia and aplastic anemia.

Bones, joints: disorder of cartilages and subchondral tissue.

Others: increasing of asthama and polypus of nose, skin rash.

Threats to Vulture survival

Diclofenac: According to Bombay Natural History Society (BNHS), veterinary use of diclofenac is the main threat to the Vultures in India. The widespread use of diclofenac as pain reliever in cattle is the cause of Vulture's mortality in India.

Habitat destruction: Developmental activities like establishment of power projects, irrigation projects, industrial units, construction of highways etc. have ruined the habitats of Vultures resulting into decline in their population.

Pesticide pollution: The chlorinated hydrocarbon D.D.T (Dichloro Diphenyl Trichloroethane) used as pesticide enters the body of Vultures through food chain where it affects the activity of estrogen hormone, as a result of which the egg shell is weakened consequently the premature hatching of egg takes place causing the death of the embryo.

Slow breeding rate: Vultures lay a single egg in a breeding season. Hence their slow breeding rate is also a threat to their survival.

Use of poisoned carcasses: Poison used by human beings to kill cattle-marauding carnivores is also a threat to Vultures in India, as consumption of such poisoned carcasses by Vulture leads to their death.

Lack of legal protection: Out of nine species of Vultures found in India only one that is the Bearded Vulture (*Gypaetus barbatus*) is protected by law and hence lack of legal protection is also a threat to their survival.

Environmental Contaminant: It has been observed that environmental contaminants cause heavy mortality in raptors and vulture population. Insecticides and pesticides contaminate the environment and their accumulation within the water bodies, in turn, serves as a potential contamination source. This may cause heavy loss to these birds due to bio-magnification.

Effect of non-steroidal anti-inflammatory drugs

The effect of NSAIDs have been widely studied and under research. Several NSAIDs available in the market out of which 11 were reported as livestock treatment (Galligan *et al.,* 2020) and up to now only five NSAIDs ketoprofen, meloxicam, carprofen, flunixin and phenylbutazone have been tested for toxicity on *Gyps* vulture *Gyps coprotheres* under controlled conditions. Out of these five NSAIDs only meloxicam has proven safe for vulture (Swarup *et al*., 2007). It was used to treat cattle's because of its lower toxicity and its two main characteristics-

1. It can be metabolised in shorter period of time and thus doesn't accumulates in the vulture's liver.
2. It has short half life (T1/2: time taken for a substance to lose one half of its pharmacological activity).

T1/2 of the meloxicam is 0.5 hours while T1/2 of diclofenac is 12 hours. It means meloxicam remains active for one hour after treatment, whereas diclofenac is still active 24 hours after treatment (Fourie *et al*., 2015).

Both diclofenac and ketoprofen have been proven fatal for vulture in wild, ketoprofen is toxic for critically endangered *Gyps* species of vulture. (Naidoo *et al.,* 2009). Other vulture species viz *Aegypius monachus* and Eurasian griffon (*Gyps fulvus*) has declined due to toxicity of diclofenac ingestion (Ahamed *et al*., 2005). A research conducted in European countries diclofenac has proven toxic to the Spanish vulture. (Daniel cressy, 2014) .Other Nsaids show lower toxicity effect.

Duck found safe when treated with ketoprofen (Mulachy *et al.,* 2003). *Gyps africans* died just after treatment of carprofen and ketoprofen. Flunixin causes 13% mortality in *Gyps* vulture (Zorilla *et al*., 2015)

In South Africa testing of carprofen used as "vulture friendly NSAIDs "in testing but it proven lethal for vulture. Carprofen ingestion doesn't show any symptoms as shown by diclofenac ingestion like neck dropping and lethargy. After carprofen ingestion- birds simply dead after 3 days of carprofen ingestion birds simply dies without pre symptoms. (Fourie *et al*., 2015).

Use of NSAIDS also affects the breeding rate or growth rate of *Gyps indicus* as reported in Rajasthan and Madhya Pradesh in between the years 2007 to 2016. (Christopher *et al*., 2021). In American kestrels up to 2 mg / kg BW meloxicam treatment, doesn't show toxicity in nephron of the bird's excretory system. (Summa *et al.,* 2017).

If the high doses of the drug meloxicam have given then it shows hepatorenal toxicity effect (i.e. nephrotoxic effect on kidney and hepatotoxic effect on liver tissues)in the poultry farms i.e. in chickens. (Hedau, 2016; Hersh *et al.,* 2005; Fletcher *et al.,* 2006; Gooch *et al.,* 2007). Up to 1 mg /kg BW meloxicam drug administration in chicken maintain the plasma concentration. (Souza MJ *et al.,* 2017)

The pharmacokinetic properties of meloxicam have been evaluated in several avian species, including chickens, African grey parrots (*Psittacus erithacus erithacus*; Montesinos *et al.,* 2017), red-tailed hawks (*Buteo jamaicensis*), and great-horned owls (*Bubo virginianus*; Lacasse *et al*., 2013).

The pharmacokinetics for ISA Brown chickens were calculated using the non-compartmental model. The elimination half-life, steady state volume of distribution and mean resident time were significantly higher in the 24 month old ISA Brown chickens compared to the 18 week old ISA Brown chickens.(Gildersleve, 2015).

A few studies have been conducted which investigate the efficacy and analgesic effects of meloxicam in avian species (Machin, 2005). Post-operative administration of meloxicam (2 mg/kg) has been shown to be the most effective in treating orthopaedic pain following an osteotomy of the left femur in pigeons (*Columba livia*) when compared to the lower doses (0.5 and 1 mg/kg) or the control group. The pigeons (*Columba livia*) given the highest dose of meloxicam showed the ability to weight bear on the affected limb better than any other group in this trial (Desmarchelier *et al.,* 2012).

Meloxicam is available in tablets, oral suspension (OA) and injectable via intravenous (IV) or intramuscular (IM) routes. Various studies have used Hispaniolan Amazon parrots (*Amazona ventralis*) to investigate the effects of meloxicam via these different administration routes (Molter *et al.,* 2013). From the study conducted by Molter and colleagues the target plasma concentration of meloxicam which is known to result in analgesia in the Hispaniolan Amazon parrots (*Amazona ventralis*) was 3.5 μg/mL. After 1 mg/kg was administered, the parrots who were injected IV or IM had plasma concentrations above this target concentration, with 3.7±2.5 and 3.5±2.2 μg/mL respectively. Furthermore, at 6 h post OA the average peak concentration was 3.5±1.2 μg/mL (Molter *et. al.,* 2013). Comparing the pharmacokinetic data of the different administration routes, the birds that were given meloxicam orally, bioavailability of meloxicam was between 49-75% compared to the bioavailability of IM or IV which were 100%. Whereas, the elimination half-life of meloxicam was similar between all 3 routes; 15.9±4.4, 15.1±7.7 and 15.8±8.6 h in IV, IM and OA respectively (Guzman, 2014).

Unlike many NSAIDs, meloxicam shows little to no adverse effects within the body after short-term use. Dijkstra and colleagues used Hispaniolan Amazon parrots (*Amazona ventralis*) to evaluate these possible side effects from blood, serum, faecal and urine samples following the administration of 1.6 mg/kg of meloxicam for 14 days. One of the main markers of renal toxicity is N-acetyl-N-D-glucosaminidase, which was examined in both urine and serum samples. Moreover, in the urine samples they screened for casts, specific gravity, protein, glucose, pH and haemoglobin, and the faecal samples were screened for blood. The serum samples were screened for: N-acetyl-N-D-glucosaminidase, glutamate dehydrogenase, alanine transaminase, aspartate aminotransferase, total protein, albumin and cholesterol also measured the whole blood clotting time. From all of the screening from the various sample types, there was no evidence to suggest that at this dosage meloxicam had any effect on the renal, gastrointestinal system or on haemostatic processes (Dijkstra *et al.,* 2015; Guzman, 2014).

Japanese quail (*Coturnix japonica*) were given meloxicam (every 12 hours for 14 days), the post-treatment uric acid concentration was significantly higher compared to the untreated quail. However, the mean value of these treated quail (7.8 ng/mL) was still within the acceptable range of uric acid concentrations. Histopathological studies of the kidneys showed no significant lesions due to the effects of meloxicam. The only adverse effect that was found during this study was muscle necrosis at the injection site. In particular, findings showed that 85% of the treated birds had haemorrhage and myositis within the pectoral muscle (injection site). The use of meloxicam is commonly used due to the ease of administration. Both oral and injectable formulations are available in the concentrations, which are suitable for small patients. However, for birds that have pre-existing renal disease or are dehydrated, it is recommended to avoid the use of any NSAIDs, including meloxicam, as this may be fatal due to the increased pressure on the kidneys (Sinclair *et al.,* 2012).

Research done on the adverse effects of meloxicam in which it may adversely affects the renal, gastrointestinal, or haemostatic systems in certain avian species (Pereira and Werther, 2007; Desmarchelier *et al.,* 2012; Dijkstra *et al.*, 2015).

If meloxicam administered at a dose of 1.6mg/kg PO every 12 hours for 15 days – has been found to be therapeutic on the basis of pharmacodynamic and pharmacokinetic data available in healthy Hispaniolan Amazon parrots, without resulting in clinical evident adverse effects, the same may not be true for other avian species or unhealthy individuals. Therefore, assessment of specific organ function may be required in some cases to ascertain suitability of meloxicam use for individual patients.

Montesinos *et al.,* (2015) concluded that no significant differences were found between hematologic and plasma biochemical values before and after 14 days of treatment with meloxicam, except for a slight increase in median a simple protein(beta globulin) and corresponding total globulin concentrations, and a small decrease in median phosphorus concentration. According to these results, meloxicam dose administered in this study protocol does not appear to cause renal disease in African grey parrots.

Dijkstra *et al.,* (2015) suggested that birds treated with meloxicam had a significant increase in number of WBCs and decrease in PCV from before to after treatment. There is significant reduction in PCV also, compared with results for the control group; however, WBC count and PCV for all birds remained within reference ranges throughout the study. There was single high value for urine NAG activity when one parrot treated with meloxicam.

An experiment study was conducted by Hedau and Bhandarkar (2016) to investigate meloxicam toxicity in poultry. Toxicity was induced by oral administration of meloxicam to birds. The observations revealed decrease in Hb, PCV, and TEC along with leucocytosis and absolute heterophillia in higher dose treated groups. Serum uric acid, creatinine, AST and ALT values were found elevated in high dose groups. The serum biochemical deviations indicated nephrotoxic and hepatotoxic effects of meloxicam in poultry at higher doses of the drug.

The effectiveness of meloxicam was observed in the treatment of proventricular dilatation disease induced by challenge with avian bornavirus (ABV), strain M24. Three groups of 4 cockatiels were employed. Group 1 was challenged with ABV alone, group 2 was challenged with ABV and 20 days later began daily oral meloxicam treatment, and group 3 received daily oral meloxicam alone. Group 2 birds died or were euthanized on or before day 98 because of severe weight loss and depression. After Necropsy and histopathology it is indicated that all birds in group 2 had pathologic evidence of proventricular dilatation disease. In all the vital organs of the group 2 birds ABV was detected by reverse transcriptase polymerase chain reaction testing. In 2 birds of group 1 there was enlarged proventriculus, and in this group 75% of the birds had ABV detectable in all their major organs. In group 3 birds No significant lesions were observed. (Hoppes *et al.*,2013).

CONCLUSION

The whole earth has its balance in several ways there are some factors which help to make the earth's atmosphere clean. These factors include microbes, mammals, insects, birds etc. Among these entire organisms in bird's vulture are called as environment's friends as they clean the environment by feeding on the dead animal and thus dispose the dead bodies avoiding harmful effects on the nature. Vulture is the scavenger and these are the important part of the ecosystem. But from last many decades its population on the earth and mainly in south Asian countries especially India, is declining rapidly. Scientist, researchers are trying find out the cause of their vanishing population. For this several studies done and different scientists and scholars found different reasons like habitat destruction, food poisoning, ecosystem unbalancing, adverse drugs introduction into body of vulture and destruction of its body organs. Main cause found is the poisoning of drugs namely diclofenac, its non steroidal anti-inflammatory drugs which is used as painkiller for veterinary purpose to treat the cattles. Other NSAIDs like caprofen, ketoprofen etc. affects these birds and other organisms of the ecosystem. As vultures feed on carrion the drug by the process of biomagnifications enters the vulture body and cause severe threat to the kidney and least affect other organs. Several programs and conservation methods are present to save vulture population. In these conservation methods some are protection, in *situ* conservation, and replace of dicofenac with its another substitutes like meloxiam as it has been proven safest drug to treat the livestock as its half life is too short.

ACKNOWLEDGEMENT

We are grateful to the department of Zoology Jai Narain Vyas University, Jodhpur Rajasthan that provided all required facilities for research work. We would like to thank Former Dean of the science faculty and professor of zoology department Dr. Ashok Purohit , who guide us for such environment conservative research work.

REFERENCES

- Ahamed, S., and Khan, U. (2005). Evaluation of current status of griffon and cinereous vultures in Balochistan Provience vulture survey.
- Ali, S. and Ripley, S.D. (1998). Handbook of the birds of India and Pakisthan. Bombay, New York, Oxford University Press,1968-1974.

- Ansari, N.M.(2015). Status and distribution of vultures in Gautam Budth Nagar District, Uttar Pradesh, India. *International Journal Of Advanced Research*. 3(5):506-511.
- Bird life International (2010). www.birdlife.org/news/news/2005/03/vulture.
- Chaudhry, J.M.I., Ogada, D.L., Malik, R.N., Virani, M.Z., Giovanni, M.D. (2012) .First evidence that populations of the critically endangered Long-billed Vulture Gyps indicus in Pakistan have increased following the ban of the toxic veterinary drug diclofenac in South Asia. *Bird Cons. Int.* 22: 389–397.
- Christopher, J.W., Clure, M.C., Brian, W. R. and Munir, Z.V. (2021). Contrasting trend in abundance of Indian vulture (*Gyps indicus*) between two study sites in neighbouring Indian states.*Frontiers in Ecology and Evolution*. doi. 10.3389/fevo.2021.629482.
- Cuthbert, R., Pain, D.J., Green, R.E., Swan, G., Swarup, D. (2006). Comparative toxicity studies of NSAIDs in birds: A criticism of Reddy *et al., Environmental Toxicology and pharmacology*. DOI:10.1016/j.etap.2006.10.001
- Cuthbert, R.J., Paudel, S., Chaudhary, A., Kaphle, K., Thapa, J., Subedi, T. R. and Taggart, M. A. (2014b) Aligning species conservation with animal welfare: formulation of vulture-safe meloxicam manufactured in South Asia and the reaction of goats to its administration. Sandy, Bedfordshire, UK: Royal Society for the Protection of Birds. (RSPB Research Report No. 520).
- Cuthbert, R. J., Taggart, M. A., Prakash, V., Chakraborty, S. S., Deori, P., Kulkarni, M.. Ranade, S., Saini. M., Sharma, A.K, Shringarpure, R. and Green, R. E. (2014a) Avian scavengers and the threat from veterinary pharmaceuticals. *Phil. Trans. Roy. Soc.* B 20130574.
- Cuthbert, R.J., Dave, R., Chakraborty, S.S., Kumar, S., Prakash, S., Ranade, S.P., Prakash,V., (2011).Assessing the ongoing threat from veterinary non-steroidal anti-inflammatory drugs to Critically Endangered *Gyps* vultures in India. *Oryx* 45(03): 420-426.
- Cressy, D. (2014). Poisoned vulture could Herald European Bird crisis. *Nature*. https://doi.org/10.1038/nature.2014.16161.
- Desmarchelier, M., Troncy, E., Fitzgerald, G., and Lair, S. (2012). Analgesic effects of meloxicam administration on postoperative orthopedic pain in domestic pigeons (*Columba livia*).*American Journal of Veterinary Research*, 73(3), 361-367.
- Dijkstra, B., Guzman, D.S.M., Gustavsen, K., Owens, S.D., Hass, C., Kass, P.H., and Paul-Murphy, J.R. (2015). Renal, gastrointestinal, and hemostatic effects of oral administration of meloxicam to Hispaniolan Amazon parrots (*Amazonaventralis*). *American Journal of Veterinary Research*, 76(4), 308-317.
- Fletcher, J.T., Graf, N., Scarman, A., Saleh,H. and Alexander, S. I. (2006)Nephrotoxicity with cyclooxygenase 2 inhibitor use in children. *Pediatric Nephrology* 21: 1893–1897.
- Fourie, T., Cromarty, D., Duncan, N., Wolter, K., and Naidoo, V. (2015) The Safety and Pharmacokinetics of Carprofen, Flunixin and Phenylbutazone in the Cape Vulture (*Gyps coprotheres*) following Oral Exposure. *Plos One* 10(10): e0141419.
- Galligan, T.H., Mallord, J.W., Prakash, V.M., Bhusal, K.P., Alam, A.S., Anthony, F.M., Dave, R., Dube, A., Shastri, K., Yogesh Kumar, Prakash, N., Ranade, S., Shringarpure, R., Chapagain, D., Chaudhary, I.P., Joshi, A.B., Paudel, K., Kabir T., Ahamed, S.,

Azmiri, K.Z., Cuthbert, R.J., Bowden, C.G.R., and Green, R.E., (2020). Trends in the availability of the vulture-toxic drug, diclofenac, and other NSAIDs in South Asia, as revealed by covert pharmacy surveys. *Bird Conservation International*, 1-17. DOI: 10.1017/S0959270920000477.

- Gilbert, M., Virani, M.Z., Watson, R.T., Oaks, J.L. , Benson, P.C., Khan, A.A., Ahmed, S., Chaudhry, J., Arshad, M., Mahmood, S., & Shah, Q.A., (2002). Breeding And Mortality Of Oriental White-Backed Vulture *Gyps bengalensis* In Punjab Province, Pakistan. *Bird Conserv. Int.* 12: 311-326.
- Gildersleve, M. (2015). Effect of Age on the Pharmacokinetics of Meloxicam in ISA Brown Chickens (*Gallus gallus domesticus*). A thesis presented in partial fulfilment of the requirements for the degree of Master of Science in Physiology at Massey University, Palmerston North, New Zealand
- Gooch, K., Culleton, B. F., Manns, B. J., Zhang, J. G., Alfonso, H., Tonelli, M., Frank, C., Klarenbach, S. and Hemmelgarn, B.R. (2007). NSAID use and progression of chronic kidney disease. *Am. J. Med.* 120(3).
- Guzman, D.S.M. (2014). Advances in avian clinical therapeutics. *Journal of Exotic Pet Medicine*, 23(1), 6-20.
- Hedauand, M. and Bhandarkar, A.G. (2016). Pathology of meloxicam toxicity in cockerel birds. *Indian Journal of Veterinary Sciences and Biotechnology*, 11(3)59-64.
- Hersh, E. V., Lally, E. T. and Moore, P. A. (2005). Update on cyclooxygenase inhibitors: has a third COX isoform entered the fray? *Current Med. Res. and Opinion* 21:1217–1226
- Hoppes, S., Heatley, J.J., Guo , J., and Turner, D., (2013). Meloxicam Treatment in Cockatiels (*Nymphicus hollandicus*) Infected With Avian Bornavirus. *Journal of Exotic Pet Medicine* 22(2013):275-279 .
- IUCN (2004). IUCN Red List of threatened Species. ,http://www.iucnredlist.org.
- IUCN (2020) Species Survival Commission Vulture updates No 13 - October 2020 - Around the World of Vultures & VSG activities.
- Lacasse, C., Gamble, K.C. and Boothe, D.M. (2013). Pharmacokinetics of a single dose of intravenous and oral meloxicam in red-tailed hawks (*Buteojamaicensis*) and great horned owls (*Bubo virginianus*), *J Avian Med Surg* 27(3): 204-210.
- Machin, K.L. (2005). Avian analgesia. Seminars in *Avian and Exotic Pet Medicine*, 14(4), 236-242.
- Molter, C. M., Court, M. H., Cole, G. A., Gagnon, D. J., Hazarika, S., and Paul-Murphy, J. R. (2013). Pharmacokinetics of meloxicam after intravenous, intramuscular, and oral administration of a single dose to Hispaniolan Amazon parrots (*Amazon aventralis*). *American Journal of Veterinary Research*, 74(3), 375-380.
- Montesinos, A., Ardiaca, M., Gilabert, J.A., Bonvehi, C., Oros, J. and Encinas,T. (2017). Pharmacokinetics of meloxicam after intravenous, intramuscular and oral administration of a single dose to African grey parrots *(Psittacus erithacus). J Vet Pharmacol Ther.* 40(3):279-284
- Montesinos, A., Ardiaca, M., Juan-Sallés, C.and Tesouro, M.A.(2015).Effects of Meloxicam on Hematologic and Plasma Biochemical Analyte Values and Results

of Histologic Examination of Kidney Biopsy Specimens of African Grey Parrots (*Psittacus erithacus*).*J Avian Med Surg* 29(1):1-8.

- Mulcahy, D.M., Larsen,S. R.,(2003).Differential mortality of male spectacled eiders (*Somateria fischeri*) and king eiders (*Somateria spectabilis*) subsequent to anesthesia with propofol, bupivacaine, and ketoprofen. *J. Avian Med. Surg.* 17, 117–123 (doi:10.1647/2001-024) .
- Naidoo, V., Wolter K., Cuthbert R., Duncan N.(2009).Veterinary diclofenac threatens Africa's endangered vulture species. *Regul. Toxicol. Pharm.* 53,205–208 .
- Naidoo, V., Taggart, M.A., Duncan,N., Wolter,K., Chipangura,J., Green,R.E., Galligan,T.H. (2018). The use of toxicokinetics and exposure studies to show that carprofen in cattle tissue could lead to secondary toxicity and death in wild vultures. *Chemosphere* ,190: 80-89.
- Naidoo, V., Wolter, K., Cromarty, D., Diekmann, M., Duncan,N., Meharg, A. *et al.*(2010). Toxicity of non-steroidal anti-inflammatory drugs to *Gyps* vultures: a new threat from ketoprofen. *Biology letters*, 6(3), 339-341.
- Pereira, M.E., and Werther, K. (2007). Evaluation of the renal effects of flunixin meglumine, ketoprofen and meloxicam in budgerigars (*Melopsittacus undulatus). Vet Rec.* ;160:844–846.
- Prakash, V. (1999). Status of vulture in Keoladeo National Park, Bharatpur, Rajasthan, with special reference to population crash in *Gyps* species. *J. Bombay Nat. Hist. Soc.,* 96: 365-368.
- Prakash, V., Pain, D. J., Cunningham, A.A., Donald, P.F., Prakash, N., Verma, A., Gargi, R., Sivakumar, S. and Rahmani, A. R. (2003) Catastrophic collapse of Indian white-backed *Gyps bengalensis* and long billed *Gyps indicus* vulture populations. *Biol. Conserv.* 109: 381–390.
- Purohit, A. and Saran, R. (2013). Population Status and Feeding Behavior of Cinereous Vulture (*Aegypus monachus*): Dynamics and Implications for the Species Conservation in and Around Jodhpur. *World Journal of Zoology*, 8 (3): 312-318,.
- Saran, R.P., Purohit, A., Heera Ram (2016).A Comparative Patho-Physiological Study of Diclofenac and Meloxicam Induced Toxicity In *Gallus domestics*. *American journal of pharmacy and health research. Volume 4, Issue* 11, ISSN: 2321–3647
- Sinclair, K. M., Church, M. E., Farver, T. B., Lowenstine, L. J., Owens, S. D., and Paul-Murphy, J. (2012).Effects of meloxicam on hematologic and plasma biochemical analysis variables and results of histologic examination of tissue specimens of Japanese quail (*Coturnix japonica*)..*American Journal of Veterinary Research*, 73(11), 1720-1727.
- Souza M.J., Bergman,B.J., White,M.S., Gordon,K.I., Gerhardt, E.L., and Cox,K.S. (2017). Pharmacokinetics and egg residues after oral administration of a single dose of meloxicam in domestic chickens (*Gallus domesticus*). *Am.J.Vet.Res* . 78(8):965-968.
- Summa, N.M., Guzman, D.S., Larrat, S., Troncy, E., Bird, D.M., Lair, S., and Fitzgerald,G.(2017).Evaluation of High Dosages of Oral Meloxicam in American Kestrels (*Falco sparverius*). *J Avian Med Surg*. 2:108-116.

- Swan, G., Naidoo, V., Cuthbert, R., Green, R. E., Pain, D. J., Swarup, D., Prakash, V., Taggart, M., Bekker, L., Das, D., Diekmann, J., Diekmann, M., Killian, E., Meharg, A., Patra, R. C., Saini, M. and Wolter, K. (2006b). Removing the threat of diclofenac to Critically Endangered Asian Vultures. *PLoS Biol.* 4: 396–402.
- Swan, G., Naidoo, V., Cuthbert, R., Green, R.E., Pain, D.J., Swarup, D., Prakash, V., Taggart, M., Bekker, L., Das, D., Diekmann, J., Diekmann, M., Killian, E.,Meharg, A., Patra, R.C., Saini, M., and Wolter, K. (2006): Removing the Threat of Diclofenac to Critically Endangered Asian Vultures. *PLoS Biol.*, 4(3): e66
- Swarup, D.P.R.C., Patra, R.C., Prakash, V., Cuthbert, R., Das, D., Avari, P., Pain, D.J., Green, R.E., Sharma, A.K., Saini, M., Taggart, M. (2007). Safety of meloxicam to critically endangered *Gyps* vultures and other scavenging birds in India. *Anim. Cons.* 10(2): 192-198.
- Zorrilla, I., Martinez, R., Taggart, M.A., and Richards, N. (2015) Suspected flunixin poisoning of a wild Eurasian Griffon Vulture from Spain.. *Cons. Biol,* 29(2), 587- 592.

CHAPTER

16

A Comprehensive Review of Non-Steroidal Anti-Inflammatory Drugs

Archana Vyas and Ashok Purohit
Department of Zoology, Jai Narain Vyas University, Jodhpur (Rajasthan), India-342001
Email: purohit1411@rediffmail.com

ABSTRACT

Inflammation is a distinguished indicator of every disease. Currently several people suffer from chronic inflammation that is caused by a variety of factors that seriously undermine health and age. NSAIDs, non-steroidal anti-inflammatory drugs, are one of the most commonly prescribed classes of medication for pain and inflammation. One of the most common action s of NSAIDs is inhibition of cyclooxygenase (COX) enzyme, which is responsible for synthesis of various prostaglandins (PGs) and thromboxanes from archidonic acid. Although NSAIDs provide great benefit in terms of pain relief, analgesic and antipyretic (fever-reducing) effects, and anti-inflammatory effects but long term use of these drugs has been associated with various adverse side effects. Thus, the present study was evaluated the action of mechanism of drug NSAIDs and their effect on the physiological fuctions.

Key Words: Inflammation, cyclooxygenase, analgesic, archidonic acid, prostaglandins.

INTRODUCTION

NSAIDs, non-steroidal anti-inflammatory drugs, are one of the most commonly prescribed classes of medication for pain and inflammation (Schjerning *et al.*, 2020). It has analgesic, antipyretic and anti-inflammatory properties, mainly used to relieve acute and chronic pain and inflammation. They are responsible for approximately 5-10% of all medications prescribed each year (Wongrakpanich *et al.,* 2018). The history of non- steroidal anti-inflammatory drugs can be traced back to ancient Egypt (an Egyptian medical text from ca. 1543 BC) where extract of willow bark was used for treatment of inflammation. These medications inhibit Cyclooxygenases (COXs) enzymes, which is responsible for synthesis of various prostaglandins (PGs) and thromboxanes from Arachidonic acid. The inhibition of COXs results in reduced Prostaglandin E2 (PGE2) levels which provides these drugs it's anti-inflammatory property (Wang *et al.*, 2021).

The worldwide NSAID market for both occasional and chronic users has been conservatively estimated at over 70 million people and more then 30 billions of non-prescription pills are purchased annually in the United States (Green, 2001). Although NSAIDs provide great benefit in terms of pain relief, analgesic and antipyretic (fever-reducing) effects, and anti-inflammatory effects but long term use of these drugs has been associated with various adverse side effects.

Pharmacodynamics and Pharmacokinetics of NSAIDs

Nonsteroidal anti-inflammatory drugs (NSAIDs) elicit their therapeutic activities through inhibition of cyclooxygenase (COX) enzyme, which is responsible for prostaglandin biosynthesis (Martel-Pelletier *et al.*, 2003). Cyclooxygenase is the enzyme that catalyzes the synthesis of cyclic endoperoxides from arachidonic acid to form prostaglandins (PGs). Generally, the NSAIDs work by inhibition of both isoforms COX-1 and COX-2. In addition to inhibition of COX, other mechanisms through which NSAIDs derive their anti-inflammatory properties are - reduction of superoxide radicals, inhibition of cell adhesion molecule expression, induction of apoptosis, decrease of nitric oxide synthase, reduction in pro inflammatory cytokine levels (tumor necrosis factor-a, interleukin-1), modification of lymphocyte activity and alteration of cellular membrane functions (Livshits and Seidman, 2010).

Arachidonic Acid Metabolism

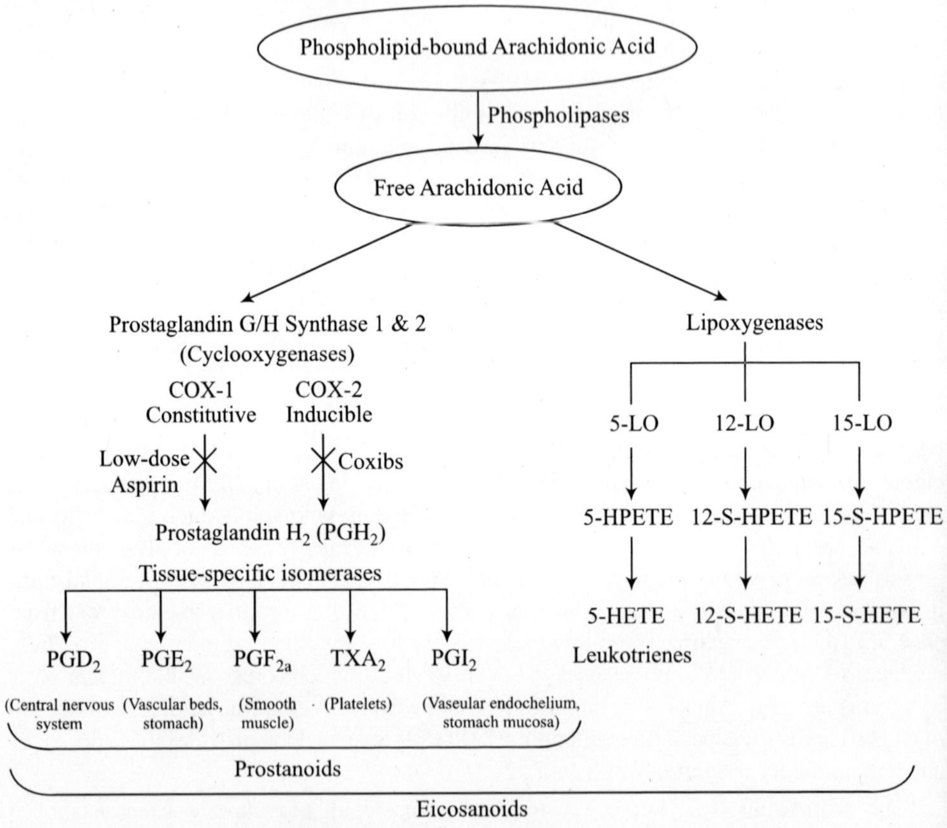

Fig. 1. *Archidonic acid metabolism: The major metabolites of arachidonic acid produced by the cyclooxygenase (cox) and lipoxygenase (LO) pathway are indicated. Example of tissue in which individual prostanoid exert prominent effects are in parentheses. PGD_2 = Prostaglandin D_2; PGE_2 = Prostaglandin E_2; PGF_{2a} = Prostaglandin F_{2a}; TXA_2 = thromboxane A_2; PGI_2 = Prostaglandin I_2; HPETE = hydroxyperoxyeicosatetraenoid acid; HETE = hydroxyeicosatetraenoid acid* (Cashman, 1996).

Generally, the NSAIDs inhibit both COX-1 and COX-2. The classical COX inhibitors, such as aspirin, ibuprofen, naproxen and flurbiprofen are not selective and inhibit all types of COX isoforms (Limongelli *et al.,* 2010). The resulting inhibition of prostaglandin and thromboxane synthesis has the effect of reduced inflammation as well as anti-pyretic, anti-thrombotic and analgesic effects. The most adverse effect of NSAIDs is induced gastrointestinal complications (Rafaniello *et al.,* 2016). NSAIDs provide most benefit by acting on the COX-2 enzyme. As a consequence, a selective inhibition of COX-2 has been sought for decades, and recently a new generation of NSAIDs, namely coxibs (i.e. Celecoxib, rofecoxib, lumiracoxib, and valecoxib), was found (Kalgutkar *et al.,* 2000). New NSAIDs are being developed that are many more times specific to the COX-2 than COX-1 enzyme.

Most NSAIDs are well absorbed in the GI tract and have high bioavailability. Some drugs such as diclofenac undergo hepatic first-pass metabolism which resulted in the reduction in bioavailability. While some drugs such as sulindac and parecoxib are prodrugs and need hepatic metabolism to become their active metabolites (sulindac sulfide and valdecoxib, respectively). NSAIDs are highly bound to plasma proteins and generally metabolized in the liver and excreted in the urine. Common NSAIDs have a variable half-life ranging from anywhere between 15-20 minutes for aspirin to 45-50 hours for piroxicam (Awtry and Loscalzo, 2000) .

Current Guideline and the Therapeutic Use of NSAIDs

NSAIDs are usually indicated for the treatment of acute or chronic conditions where pain, fever and inflammation are present. They are commonly used for treating musculoskeletal disorder (chronic, like osteoarthritis and rheumatoid arthritis (Bournia *et al.,* 2016), and acute, like acute gout and injuries), Inflammatory arthropathies (e.g. Ankylosing spondylitis, Psoriatic arthritis, Reactive arthritis), metastatic bone pain, Mild-to-moderate pain due to inflammation and tissue injury, headaches, migraine, Pyrexia (fever), Renal colic, dental and postoperative pain and dysmenorrhea. Aspirin, the only NSAID able to irreversibly inhibit COX-1, is also indicated for inhibition of platelet aggregation (Vyas *et al.,* 2016). Research continues into their potential for prevention of colorectal cancer, and treatment of other conditions, such as cancer and cardiovascular disease.

Professional societies, including American Geriatric Society, American College of Rheumatology, and the European League Against Rheumatism, recommend using NSAIDs with caution and limit their use to the lowest effective dose and shortest duration. Unfortunately, NSAIDs are commonly associated with a wide range of toxic side effects. Long term use of NSAIDs cause several serious adverse effects. Use of the majority of NSAIDs increases with age, primarily for symptoms associated with osteoarthritis and other chronic musculoskeletal conditions (Griffin, 1998). The most prominent members of this group of drugs are aspirin, ibuprofen, naproxen and diclofenac, all of which are available over the counter in most countries.

Alzheimer's Disease (AD) Lowering Effect of NSAIDs

Alzheimer's disease (AD) is the most common form of dementia. it is an age-associated neurodegenerative disease that is progressive in nature, starting slow and worsening over time (Moore *et al.,* 2010). According to some estimates AD affects roughly 5% of the population over the age of 65 (Imbimbo *et al.*, 2010). Many epidemiological studies have suggested that long term use of non-steroidal anti-inflammatory drug (NSAID) like Sulindac, Nabumetone, Ketoprofen, Piroxicam, Naproxen, Diclofenac, Paracetamol and Aspirin, Ibuprofen and Naproxen all have a role in decreasing the risk for AD progression by reducing brain Aβ levels, amyloid plaque burden, and lowering microglial activation (Nevado-Holgado and Lovestone, 2017).

Rheumatoid Arthritis and NSAIDs

Arthritis is a chronic inflammatory autoimmune disease, which is associated with considerable pain and inflammation – typically around joins, and has a significant impact on patient's functional capacity and overall quality of life (Moore *et al.*, 2005). While Non-steroidal anti-inflammatory drugs (NSAIDs) aren't used to stop the progression of the disease, they play an important role in managing symptoms and relieving pain due to their anti-inflammatory and analgesic properties. NSAIDs are widely used in combination of other drugs for the treatment of rheumatoid arthritis as it reduces inflammation which helps to relieve joint pain and stiffness (Crofford, 2013).

Anticancer Activity of NSAIDs

NSAIDs may have value as effective chemo preventive agents. Recent epidemiological studies and clinical trials suggest that long-term use of aspirin and other non-steroidal anti-inflammatory drugs (NSAIDs) are beneficial in reducing the risk of cancer. They have been observed to show decrease in the incidence of certain malignancies, including colorectal, oesophageal, breast, lung, prostate and bladder cancers (Nelson and Harris, 2000). Colorectal cancer is the 3rd most common malignant neoplasm worldwide (Piazza *et al.*, 2010; Stolfi *et al.*, 2013). Long-term use of NSAIDs can significantly decrease the risk of death from colorectal cancer. In the tumor microenvironment PGE_2 is supposed to play a key role in cancer progression, it promotes tumor growth and metastasis (Wang and DuBois, 2018). Aspirin and other NSAIDs have regulatory effects on the gene transcription and protein synthesis of multiple molecules involved in several inflammatory and neoplastic pathways (Stolfi *et al.*, 2013).

Atherosclerosis and NSAIDs

Atherosclerosis, the fundamental cause of coronary heart disease and stroke, is a complex chronic inflammatory disease (Purohit and Ram, 2012). Cyclooxygenase is a key enzyme for inflammation. It regulates monocytes, macrophages, endothelial cells and smooth muscle cells in atherosclerotic lesions. NSAIDs like COX inhibitors and COX-2 selective inhibitors might have protective effects within the cardiovascular system or even reduce atherogenesis by their anti-inflammatory property. The non-selective NSAIDs like indomethacin, aspirin, or sulindac also inhibits neointimal formation after arterial injury (Metzner *et al.*, 2007). The beneficial effects of low dose of aspirin in cardiovascular event are due to its ability to inhibit platelet aggregation through inhibition of COX-1 mediated production of platelet TXA_2 (Linton and Fazio, 2008). Aspirin also helps in reducing the progression of coronary atherosclerosis.

NSAIDs' Known Adverse Effects

NSAIDs are considered group one medications, recommended for mild pain and are the first step in treating pain of acute and chronic painful and inflammatory conditions, the long-term use of these drugs induces several side effects. Common side effects that occur with NSAIDs include:

NSAIDs and Gastrointestinal (GI) Adverse Effects

NSAIDs increase the risk of both upper and lower GI complications (Domper Arnal *et al.,* 2018). NSAIDs cause a dual assault on the GI tract: the acidic molecules directly irritate the gastric mucosa, and inhibition of COX-1 and COX-2 reduces the levels of protective prostaglandins. Inhibition of prostaglandin synthesis in the GI tract causes increased gastric acid secretion, diminished bicarbonate secretion, diminished mucus secretion and diminished trophic effects on epithelial mucosa. The fundamental mechanism of NSAIDs induced GI side effects lies in the fact that these medications inhibit prostaglandin synthesis, causing weakening of the protective GI mucosal barrier, predisposing one to bleeding (Wongrakpanich *et al.,* 2018). Clinical NSAID ulcers are associated with the systemic effects of NSAID administration. Such damage occurs irrespective of the route of administration of the NSAID (e.g., oral, rectal, or parenteral) and may occur even in patients with achlorhydria.

Ulceration risk increases with therapy duration and with higher doses. Recent studies show that over 50% of patients that take NSAIDs have sustained some mucosal damage to their intestine (Higuchi *et al.*, 2009). Elderly individuals (aged over 65 years), heavy smokers, patients who are taking other medications at the same time, patients who have a history of gastrointestinal problems, take NSAIDs in high doses, or for long-term, or those with some other conditions, such as hypertension, cardiovascular disease, or diabetes are the groups of people that may be at higher risk of developing gastrointestinal problems while taking NSAIDs.

NSAIDs and Kidney Adverse Effects

NSAIDs may lead to renal vasoconstriction and consequently reduced renal perfusion and aberrant renal function by inhibition of prostaglandin and thromboxane synthesis. Clinical manifestations of NSAID produced nephrotoxicity includes electrolyte imbalance such as hyperkalemia, reduce glomerular filtration rate, nephrotic syndrome related to drug induced minimal change disease, edema, renal failure, acute interstitial nephritis, chronic kidney disease, sodium retention, and renal papillary necrosis (*Harirforoosh et al.*, 2013). Long term use of NSAIDs are associated with adverse renal effects caused by the reduction in synthesis of renal prostaglandins.

NSAIDs and Liver Adverse Effects

Liver damage occurs infrequently, and is usually reversible. Patients with osteoarthritis develop symptomatic liver disease more often than patients with rheumatoid arthritis. Long-term treatment with NSAIDs increases the chances of serious hepatic injury (Kim *et al.*, 2021). However the impact of hepatotoxicity is not severe as GI toxicity and nephrotoxicity.

NSAIDs and Cardiovascular Adverse Effects

NSAIDs except from aspirin, both newer selective COX-2 inhibitors and traditional anti-inflammatories, increase the chance of myocardial infarction and stroke (Kearney *et al.*, 2006). They are not recommended in those who have had a previous heart attack as they increase the danger of death and / or recurrent MI (Schjerning Olsen *et al.*, 2011). NSAIDs apart from (low-dose) aspirin are related to a doubled risk of symptomatic heart failure in patients without a history of cardiac disease. In patients with such a history, however, use of NSAIDs (aside from low-dose aspirin) was related to over10-fold increase in heart failure (Page and Henry, 2000). If this link is proven causal, researchers estimate that NSAIDs would be responsible for up to twenty percent of hospital admissions for congestive heart failure. In people with heart failure, NSAIDs increase mortality risk by approximately 1.2-1.3 for naproxen and ibuprofen, 1.7 for rofecoxib and celecoxib, and 2.1 for diclofenac (Gislason *et al.*, 2009).

NSAIDs and Combinational Risk

If a COX-2 inhibitor is taken, one should not use a traditional NSAID (prescription or over-the-counter) concomitantly. In addition, people on daily aspirin therapy (e.g., for reducing cardiovascular risk) must be careful if they also use other NSAIDs, as the latter may block the cardioprotective effects of aspirin.

NSAIDs and Inflammatory Bowel Disease

NSAIDs should be used with caution in individuals with inflammatory bowel disease (e.g., Crohn's disease or ulcerative colitis) due to their tendency to cause gastric bleeding and form ulceration in the gastric lining. Pain relievers such as paracetamol (also known as acetaminophen) or drugs containing codeine (which slows down bowel activity) are safer medications for pain relief in IBD (Long *et al.*, 20016).

NSAIDs and Reproductive Organ

NSAIDs like diclofenac, aspirin etc. causes significant decrease in sperm density in testis and cauda epididymis and sperm motility as well as significant increase in the percentage of morphologically abnormal sperm cells in the treated rats (Oyedeji *et al.*, 2013; Vyas *et al.*, 2016, 2018) and also caused significant histopathological variations in reproductive organs such as ovary and uterus by modulating certain metabolism enzyme in female albino rats and in testis in male albino rats (Jain *et al.*,2012a).

NSAIDs and Erectile Dysfunction Risk

Long term use of NSAIDs, including ibuprofen was linked to 1.4 times increased risk of erectile dysfunction (Shiri *et al.*, 2006; Patel *et al.*, 2016). The report by Kaiser Permanente and published in the Journal of Urology, considered that "regular non-steroidal anti-inflammatory drug use is associated with erectile dysfunction beyond what would be expected due to age and other condition". The director of research for Kaiser Permanente added that "There are many proven benefits of non-steroidal in preventing heart disease and for other conditions.

NSAIDs and Pregnancy

Medical professionals have advised against the use of NSAIDs during pregnancy, especially during the third trimester, they may cause premature closure of the fetal ductus arteriosus and renal ADRs in the fetus. They have also been linked to premature birth (Østensen and Skomsvoll, 2004) and miscarriage (Nakhai *et al.*, 2011). Aspirin, however, is used together with heparin in pregnant women with antiphospholipid antibodies (Cervera and Balasch, 2004). On the other hand, paracetamol (acetaminophen) is considered relatively safe during pregnancy, however, a study released by Leffers et al in 2010 indicated an association between its use and male infertility in the unborn (Graham *et al.*, 2005; Kristensen *et al.*, 2011). The use of NSAIDS in expectant mothers may also make the labor process slower and last longer. Thus the doses should be taken exactly as prescribed (Wilkes *et al.,* 2005).

NSAIDs and Photosensitivity

Photosensitivity is not an effect that is usually at the forefront of anyone's mind while considering the adverse effect of the NSAIDs (Moore, 2002). However, the 2-arylpropionic acids have been shown to be the most common to produce photosensitivity reactions, with other NSAIDs including piroxicam, diclofenac, and benzydamine also showing similar effects.

Other

Other than the reactions listed above, the common adverse drug reactions (ADRs) include: raised liver enzymes, headache, and dizziness. While some uncommon ADRs include hyperkalemia, confusion, bronchospasm, rashes, and rapid and severe swelling of the face and/or body. In rare cases Ibuprofen may also cause the symptoms of irritable bowel syndrome. Some of the reported side effects of NSAIDs are also related to mental health. These effects include depression, nightmares, irritability, anxiety, and psychotic reactions (Onder *et al.*, 2004). NSAIDs have also been implicated in some instances of Stevens–Johnson syndrome. While penetration of central nervous system (CNS) by most NSAIDs is very limited, the COX enzymes that NSAIDs inhibit, are expressed constitutively in some areas of the CNS, which means that even very low levels of penetration can cause some adverse effects like as somnolence and dizziness. In some very rare instances, use of ibuprofen may result in aseptic meningitis.

Allergies to NSAIDs might exist just like any other drugs and while some of those allergies are specific to one NSAID or other, up to 1 in 5 people may also have unpredictable cross-reactive allergic responses to other NSAIDs as well (Woessner and Castells, 2003). The use of NSAIDs may result in worsening of symptoms in asthma patients hence; it is recommended that asthma patients avoid them. Also use of NSAIDs in patients after surgery or a traumatic injury may increase the risk of excessive bleeding.

Ecological Effects

The presence of pharmaceutical pollutants in both marine and freshwater sources has been of concern to the biologist studying the impact of such pollution on aquatic animals. NSAIDs dominate quantitatively among the most commonly used drugs and hence it is of vital importance to study the impact of NSAIDs in low environmental concentrations, over the long term, on aquatic life. Studies have shown not only behavioral and physiological impact but also genetic alterations or reproduction disorders among aquatic life. While the number of such studies has increased a lot in recent years, our knowledge on the impact of NSAIDs in the environment on animal populations still remains limited to certain model species understudies and the vast majority of organisms are yet to be explored in this regard (Świacka *et al.*, 2021). Hussain *et al.*, (2008) have reported toxico-pathological effects of diclofenac in different avian species (broiler chicks, pigeons, Japanese quail and mynah) as a result, severity of clinical disease has increased in a dose-related manner. Avian species showed toxic effect in liver and kidney. Diclofenac is toxic to vultures even in small doses, causing kidney failure. That results in uric acid accumulating in the birds› blood and crystallizing around their internal organs a condition called visceral gout. (Becker, 2016)

CONCLUSION

In order to provide comprehensive study, knowing the mechanism of action, current guidelines, adverse drug reaction, and the pleiotropic effects of common drugs is important. NSAIDs are one of the most commonly given drugs in term of pain and inflammation. Long term administration of NSAIDs induces dose dependent hepatotoxicity, nephrotoxicity, GI toxicity, reproductive toxicity and other side effects. These medications should be prescribed for the shortest duration possible in the lowest effective dose and with careful observation to monitor GI, renal, and cardiovascular and other toxicity. Although NSAIDs provide great benefit in terms of pain, fever and inflammation but long term use of these drugs has been associated with various adverse side effects. Thus, these risks and benefits should be balanced judiciously in each patient to optimize overall outcomes.

REFERENCES

- Awtry, E.H. and Loscalzo, J. (2000): Aspirin. *Circulation*. 101: 1206-1218.
- Becker, R. (2016): Modelling study paints bleak picture for Europe's bird populations. *Nature*. 1-2.
- Bournia, V., Kitas, G., Protogerou, A.D. and Sfikakis, P.P. (2017): Impact of non-steroidal anti-inflammatory drugs on cardiovascular risk: Is it the same in osteoarthritis and rheumatoid arthritis? *Mod. Rheumatol.* 27(4):559-569.
- Cashman, J.N. (1996): The mechanisms of action of NSAIDs in analgesia. *Drugs*. 52(5):13-23.
- Cervera, R. and Balasch, J. (2004): The management of pregnant patients with antiphospholipid syndrome. *Lupus*. 13(9):683-687.
- Crofford, L.J. (2013): Use of NSAIDs in treating patients with arthritis. Arthritis. Res. Ther., 15(3):S2.
- Domper Arnal, M., García-Rayado, G. and Lanas, A. (2018): NSAID induced gastrointestinal damage and designing GI-sparing NSAIDs. *Expert Rev. Clin. Pharmacol.* 11(10).

- *Gislason, G.H., Rasmussen, J.N., Abildstrom, S.Z., Schramm, T.K., Hansen, M.L.,* Fosbøl, E.L., et al. (2009): Increased mortality and cardiovascular morbidity associated with use of nonsteroidal anti-inflammatory drugs in chronic heart failure. *Ann. Intern. Med.* 169(2):141-149.
- Gotzsche, Pc. (1989): Methodology and overt and hidden bias in reports of 196 double-blind trials of nonsteroidal antiinflammatory drugs in rheumatoid arthritis. *Control. Clin. Trials*. 10 (1):31-56.
- Graham, G.G., Scott, K.F. and Day, R.O. (2005): Tolerability of paracetamol. *Drug safety*. 28 (3):227-240.
- Green, G. (2001): Understanding NSAIDs: From aspirin to COX-2. *Clin cornerstone*, 3(5):50-59.
- Griffin, M.R. (1998): Epidemiology of nonsteroidal anti-inflammatory drug-associated gastrointestinal injury. *Am j med.* 104(3A):23S-29S.
- *Harirforoosh, S., Asghar, W. and Jamali, F.* (2013). Advers effects of nonsteroidal antiinflammatory drugs: an update of gastrointestinal, cardiovascular and renal complications. *J Pharm Pharm Sci.* 16:821-847.
- Higuchi, K., Umegaki, E., Watanabe, T., Yoda, Y., Morita, E., Murano, M., Tokioka, S. and Arakawa, T. (2009): Present status and strategy of NSAIDs-induced small bowel injury. *J. Gastroenterol.* 44 (9):879–88.
- Hussain, I.M., Khan, Z., Khan, A., Javed I. and Saleemi, M.K. (2008): Toxicological effects of diclofenac in four avian species. *Avian Patho.,* 37(3):315- 321.
- Imbimbo, B.P., Solfrizzi, V. and Panza, F. (2010): Are NSAIDs Useful to Treat Alzheimer's Disease or Mild Cognitive Impairment? *Front Aging Neurosci.,* 2:19.
- Jain, N., Shrivastava, R., Raghuwanshi, A.K. and Shrivastava, V.K. (2012a): The effect of aspirin on reproductive organs of female albino rat. *Int. J. Pharma. Sci. Rev. Res.,* 3(8):2644-2647.
- Jain, N., Shrivastava, R., Raghuwanshi, A.K. and Shrivastava, V.K. (2012b): Aspirin induced change in serum ACP, ALP, GOT, GPT, bilirubin and creatinine in correlation with histopathological change in liver and kidney of female albino rat. *Int. J. Pharma. Sci. Rev. Res.,* 4(3):9-11.
- Kalgutkar, A.S., Crews, B.C., Rowlinson, S.W., Rowlinson, A.B., Kozak, K.R., Remmel, R.P. and Marnett, L.J. (2000): Biochemically based design of cyclooxygenase-2 (COX-2) inhibitors: Facile conversion of nonsteroidal antiinflammatory drugs to potent and highly selective COX-2 inhibitors. *Proc. Natl. Acad. Sci. U.S.A.* 97(2):925-930.
- Kearney, P.M., Baigent, C., Godwin, J., Halls, H., Emberson, J.R. and Patrono, C. (2006): Do selective cyclo-oxygenase-2 inhibitors and traditional non-steroidal anti-inflammatory drugs increase the risk of atherothrombosis? Meta-analysis of randomised trials. *BMJ.* 332(7553):1302-1308.
- Kim, M., Lee, E.J. and Lim, K. (2021): Ibuprofen Increases the Hepatotoxicity of Ethanol through Potentiating Oxidative Stress. *Biomol Ther (Seoul).* 29(2):205–210.
- Kristensen, D.M., Hass, U., Lesné, L., Lottrup, G., Jacobsen, P.R., Desdoits-Lethimonier, C., Boberg, J., Petersen, J.H., Toppari, J., Jensen, T.K., Brunak, S., Skakkebæk, N. E.,

Nellemann, C., Main, K.M., Jégou, B. and Leffers, H. (2011): Intrauterine exposure to mild analgesics is a risk factor for development of male reproductive disorders in human and rat. *Hum Reprod.* 26 (1):235-244.

- Limongelli, V., Bonomi, M., Marinelli L., Gervasio, F.L., Cavalli, A., Novellino, E. and Parrinello, M. (2010): Molecular basis of cyclooxygenase enzymes (COXs) selective inhibition. *Proc. Natl. Acad. Sci. U.S.A.* 107(12):5411-5416.
- Livshits, A. and Seidman, D.S. (2010): Role of Non-Steroidal Anti-Inflammatory Drugs in Gynecology. *Pharmaceuticals.* 3:2082-2089.
- Long, M.D., Kappelman, M.D., Martin, C.F., Chen, W., Anton, K. and Sandler, R.S. (2016): Role of Non-Steroidal Anti-Inflammatory Drugs in Exacerbations of Inflammatory Bowel Disease. *J Clin Gastroenterol.* 50(2):152-156.
- Martel-Pelletier, J., Lajeunesse, D., Reboul, P. and Pelletier, J-P. (2003): Therapeutic role of dual inhibitors of 5-LOX and COX, selective and non-selective non-steroidal anti-inflammatory drugs. *Ann Rheum Dis.* 62:501-509.
- Metzner, J., Popp, L., Marian, C., Schmidt, R., Manderscheid, C., Renne, C., Fisslthaler, B., Fleming, I., Busse, R., Geisslinger, G. and Niederberger, E. (2007): The effects of COX-2 selective and non-selective NSAIDs on the initiation and progression of atherosclerosis in ApoE−/− mice. *Int. J. Mol. Med.,* 85(6):623-633.
- Moore, A.H., Bigbee, M.J., Boynton, G.E., Wakeham, C.M., Rosenheim, H.M., Staral, C.J., Morrissey, J.L. and Hund, A.K. (2010): Non-Steroidal Anti- Inflammatory Drugs in Alzheimer's Disease and Parkinson's Disease: Reconsidering the Role of Neuroinflammation. *Pharmaceuticals (Basel).,* 3(6):1812-1841.
- Moore, D.E. (2002): Drug-induced cutaneous photosensitivity: incidence, mechanism, prevention and management. *Drug Safety.* 25 (5):345-372.
- Moore, R.A., Derry, S., Makinson, G.T. and McQuay, H.J. (2005): Tolerability and adverse events in clinical trials of celecoxib in osteoarthritis and rheumatoid arthritis: systematic review and meta-analysis of information from company clinical trial reports. *Arthritis Res. Ther.,* 8:401.
- Nakhai-Pour, H.R., Broy, P., Sheehy, O. and Bérard, A. (2011): Use of non-aspirin nonsteroidal anti-inflammatory drugs during pregnancy and the risk of spontaneous abortion. *CMAJ.* 183 (15):1713-1720.
- Nelson, J.E. and Harris R.E. (2000): Inverse association of prostate cancer and non-steroidal anti-inflammatory drugs (NSAIDs): results of a case-control study. *Oncology.,* 7(1):169-239.
- Nevado-Holgado, A. J. and Lovestone, S. (2017): Determining the Molecular Pathways Underlying the Protective Effect of Non-Steroidal Anti-Inflammatory Drugs for Alzheimer's Disease: A Bioinformatics Approach. *Comput. Struct. Biotechnol. J.,* 15:1-7.
- Onder, G., Pellicciotti, F., Gambassi, G. and Bernabei, R. (2004): NSAID-related psychiatric adverse events: who is at risk? *Drugs.* 64:2619-2627.
- Ostensen, M.E. and Skomsvoll, J.F. (2004): Anti-inflammatory pharmacotherapy during pregnancy. *Expert opin pharmacother.* 5(3):571-580

- Oyedeji, K.O., Bolarinwa, A.F. and Adigun, A.K. (2013): Effect of Aspirin on Reproductive Functions in Male Albino Rats. *IOSR. J. Pharm. Bio. Sci.,* 4:49-54.
- Page, J. and Henry, D. (2000): Consumption of NSAIDs and the development of congestive heart failure in elderly patients: an underrecognized public health problem. *Arch. Intern. Med.* 160 (6):777-784.
- Patel, D.P., Schenk, J.M., Darke, A., Myers, J.B., Brant, W.O. and Hotaling, J.M. (2016): Non-steroidal anti-inflammatory drug use not associated with erectile dysfunction risk: results from the prostate cancer prevention trial. *BJU Int.* 117(3):500-506.
- Piazza, G.A., Keeton, A.B., Tinsley, H.N., Whitt, J.D., Gary, B.D., Mathew, B., Singh, R., Grizzle, W.E., and Reynolds, R.C. (2010): NSAIDs: Old Drugs Reveal New Anticancer Targets. *Pharmaceuticals.,* 3:1652-1667.
- Purohit, A. and Ram, H. (2012): Hypolipidemic and antiatherosclerotic effects of *Prosopis cineraria* bark extract in experimentally induced hyperlipidemic rabbits. *Asian J. Pharm. Clin. Res.,* 5(3):106-109.
- Rafaniello, C., Ferrajolo, C., Sullo, M.G., Sessa, M., Sportiello, L., Balzano, A., Manguso, F., Aiezza, M.L., Rossi, F., Scarpignato, C., and Capuano, A. (2016): *Risk of gastrointestinal mplications associated to NSAIDs, low-dose aspirin and their combinations: results of a pharmacovigilance reporting system. Pharmacol. Res.*104:108-114.
- Schjerning Olsen, A.M., Fosbøl, E.L., Lindhardsen, J., Folke, F., Charlot, M., Selmer, C., Lamberts, M., Olesen, J.B., Køber, L., Hansen, P.R., Torp-Pedersen, C. and Gislason, G.H (2011): Duration of treatment with nonsteroidal anti-inflammatory drugs and impact on risk of death and recurrent myocardial infarction in patients with prior myocardial infarction: a nationwide cohort study. *Circulation.* 123 (20):2226-2235.
- Schjerning, A., McGettigan, P. and Gislason, G. (2020): Cardiovascular effects and safety of (non-aspirin) NSAIDs. *Nat. Rev. Cardiol.* 17:574–584.
- Shiri, R., Koskimäki, J., Häkkinen. J., Tammela, T.L.J., Auvinen, A. and Hakama, M. (2006): Effect of Nonsteroidal Anti-Inflammatory Drug Use on the Incidence of Erectile Dysfunction. *The Journal of urology.* 175(5):1812-1816.
- Singh, G. and Triadafilopoulos, G. (1999): Epidemiology of NSAID induced gastrointestinal complications. *J Rheumatol Suppl.* 56:18-24.
- Stolfi, C., Simone, V.D., Pallone, F. and Monteleone, G. (2013): Mechanisms of Action of Non-Steroidal Anti-Inflammatory Drugs (NSAIDs) and Mesalazine in the Chemoprevention of Colorectal Cancer. *Int. J. Mol. Sci.,* 14 (9):17972-17985.
- Świacka, K., Michnowska, A., Maculewicz, J., Caban, M. and Smolarz, K. (2021): Toxic effects of NSAIDs in non-target species: A review from the perspective of the aquatic environment. *Environ. pollut.* 273:115891.
- Vyas, A., Purohit, A. and Ram, H. (2018): Assessment of dose-dependent reproductive toxicity of diclofenac sodium in male rats. *Drug Chem. Toxicol.* 478-486.
- Vyas, A., Ram, H., Purohit, A. and Jatwa, R. (2016): Adverse Effects of Subchronic Dose of Aspirin on Reproductive Profile of Male Rats. *J Pharma.* 2016:9.

- Wang, B., Wu, L., Chen, J., Dong, L., Chen, C., Wen, Z., Hu, J., Fleming, I. and Wang, D.W. (2021): Metabolism pathways of arachidonic acids: mechanisms and potential therapeutic targets. Curr. *Signal Transduct. Ther,* 6(94):1-30.
- Wang, D. and DuBois, R.N. (2018): Prostaglandins and Cancer. *Gut.,* 55:115-122.
- Wilkes, J.M., Clark, L.E. and Herrera, J.L. (2005): Acetaminophen overdose in pregnancy. *South. Med. J.* 98 (11):1118-1122.
- Woessner, K.M. and Castells, M. (2013): NSAID single-drug-induced reactions. *Immunol Allergy Clin North Am.* 33 (2):237-249.
- Wongrakpanich, S., Wongrakpanich, A., Melhado, K. and Rangaswami, J. (2018): A comprehensive review of non-steroidal anti-inflammatory drug use in the elderly. *Aging dis.* 9(1):143–150.